THZ IDENTIFICATION FOR DEFENSE AND SECURITY PURPOSES

Identifying Materials, Substances, and Items

THZ IDENTIFICATION FOR DEFENSE AND SECURITY PURPOSES

Identifying Materials, Substances, and Items

ANDRE U. SOKOLNIKOV

President, Visual Solutions & Applications, USA

World Scientific

NEW JERSEY • LONDON • SINGAPORE • BEIJING • SHANGHAI • HONG KONG • TAIPEI • CHENNAI

Published by

World Scientific Publishing Co. Pte. Ltd.

5 Toh Tuck Link, Singapore 596224

USA office: 27 Warren Street, Suite 401-402, Hackensack, NJ 07601

UK office: 57 Shelton Street, Covent Garden, London WC2H 9HE

Library of Congress Cataloging-in-Publication Data
Sokolnikov, Andre.
  THz identification for defense and security purposes : Identifying material, substances, and items / Dr. Andre Sokolnikov, Visual Solutions & Applications, USA.
    pages cm
  Includes bibliographical references and index.
  ISBN 9789814452380
  1. Terahertz technology. 2. Materials--Analysis. I. Title. II. Title: Terahertz identification for defense and security purposes of materials, substances, and items.
  TK7877.S65 2013
  621.381'3--dc23

                                                      2012046734

British Library Cataloguing-in-Publication Data
A catalogue record for this book is available from the British Library.

In-house Editor: Amanda Yun

Typeset by Stallion Press
Email: enquiries@stallionpress.com

Preface

This book came from real life.

I started working with electronic applications for defense back in the 70's. In the 80's, my senior project was about fiber-optics. It was mostly focused on optical measurements but implied a military intercom usage. In the early 90's, microwave range was widely used for surveillance and identification purposes. The first surveillance system I designed worked at 1 GHz. It was still three orders of magnitude lower than a typical terahertz (THz) frequency. Ever since then, the frequency has been increasing, filling the gap between microwaves and infrared. THz range became popular in the middle of the 90's. A lot of effort has been invested in making this range workable but there are still many challenges (mostly of technological nature). However, in a number of aspects, the scope of the scientific and engineering approaches has remained the same as it was for the microwave identification.

As we have been doing for 20 years, now, it is important to continue to use mathematical, statistical and probabilistic methods to determine what the identifiable object is (which is especially critical for automatic systems). Long gone are the days when the only reliable detector and analyzer was the human eye. Now, it is usually a microprocessor or a more elaborate computer system that makes a decision. In addition, the basic physics of the electromagnetic processes (since THz waves are part of the electromagnetic spectrum) are the same as they were for microwaves, visible light and X-rays — traditional means of identification and control. Bearing in mind the above, I tried to present succinctly the approaches applicable to the well-established means of identification. Such are the chapters concerning the electromagnetic theory and mathematical methods.

I have met many people with B.S. and advanced degrees who, for example, asked me to remind them of some electromagnetism laws even though they had taken corresponding courses in the past. Also, mathematical approaches were something that we would discuss. At the same time, I discovered that many specialists would collaborate with other experts (in the fields that they did not know) in order to make an electronic system such as a THz detection or surveillance device. I met professors from e.g. mechanical engineering departments who knew quite well how to build a mathematical model of a process, had some knowledge of electronics but needed other professors (say from Electrical Engineering) to work with image or signal processing. Thus, I would expect many potential readers to use certain chapters only as a reference or to acquire some idea of how, for example, a typical mathematical model is built and then use this knowledge to collaborate with specialists in this field who know the subject in more detail. Researchers who know the physics of THz sources and detectors might skip the corresponding chapters or just browse them. However, Ch. 7 on "THz Applications" is meant for everyone and could be useful even for those who are not directly involved with Defense and Security technology.

Having spent almost a decade working for a federal organization dealing with organized crime and terrorism, I have adopted their perspective and approaches. Since the book is devoted to defense and security issues, I have tried to render, wherever it was possible, views of the law enforcement and military agencies and their understanding of the problem. For the most part, Ch. 7 contains a number of requirements emphasized by the military agencies. There are several examples in the chapter that illustrate the common approaches and achieved level of the THz technology as well as practical implementations (some of them of my own design).

In many ways, this book is written for the readers with engineering and scientific degrees who would be interested to get involved with the design and practical exploitation of THz equipment and work with representatives of the law enforcement and defense agencies and companies. Also, I can envision an opportunity for a college or university student to get a flavor of the possibilities of THz range and make it their field of study as a result of reading this book. Lastly, I had in mind businessmen who sell, buy, and service or market THz technology and want to know something about the products that they are dealing with.

On the whole, I would expect from the readers some technical and scientific background, university-level in physics, mathematics, material science and even biology, but not necessarily a thorough grasp of every subject in question.

Andre Sokolnikov

Contents

Chapter 1

Introduction

In physics, terahertz radiation refers to electromagnetic waves with frequencies approximately between 300 gigahertz (3×10^{11} Hz) and 3 terahertz (3×10^{12} Hz), or the wavelength range between 1 millimeter (high frequency edge of the microwave band) and $100 \, \mu m$ (long-wavelength edge of far infrared light). It is so called terahertz range (THz range). Similar to infrared radiation and microwaves, these waves usually travel in line of sight. THz radiation is non-ionizing sub millimeter microwave radiation and shares with microwaves the capability of penetration of a wide variety of non-conducting materials. THz can pass through wood, paper, masonry, fabrics, plastics and ceramics. It cannot penetrate metals or water. THz radiation interacts with the molecules of the materials in question. The result of this interaction may be detected and used for identification of specific layers, structures, substances, etc. that THz radiation encounters. This offers the possibility to combine spectral identification with imaging.

Another feature-merit of THz waves is that passive detection of the THz signatures helps avoid the bodily privacy concerns of other detection methods (*e.g.*, X-rays) by being focused to a very specific range of materials and objects. THz radiation is non-ionizing, and therefore does not damage biological tissues and DNA, unlike X-rays. THz radiation can also detect differences in water content and densities of tissues as well as create 3-D imaging of biological structures (teeth, for example), thus being useful for conventional medicine, biological studies or forensics.

Applications of THz in these areas have attracted much interest: astronomy, spectroscopy, chemistry, biochemistry, manufacturing and even high-altitude telecommunications. Although penetration through the air is limited (close to the surface of the earth) that creates difficulties for distant sensing or communication purposes, THz radiation is well known in astronomy and it is possible to detect such radiation coming from space.

The last gap in the electromagnetic spectrum was filled in 1923, when Ernest Nickols and J.D. Tear succeeded in their attempts in observing radiation in the THz range. They systematically approached this problem going, both upwards in frequency from the microwave region and downward from the infrared range. However, there had been other significant contributions to the understanding of this phenomenon. One of them was Max Planks' equation that would become Plank's Radiation Law.

Initially, applications of optical gratings, interference, refraction, etc. were only partially successful at THz frequencies, and radiometers were not sensitive enough. As a result, extensive research was conducted to analyse data indirectly at harmonies that fell into the infrared range where measurements could readily be made. Computational methods were applied to extract the fundamental frequency from the observed higher-order frequencies. These measurements were made manually until 1950's, when computers could finally perform a Fourier transform on the spectral data. Up until the 1980's the use of electromagnetic waves in the far-infrared or terahertz region of the spectrum was limited due to the low intensity of thermal sources and the insufficient sensitivity of most detectors. This obstacle was removed with the introduction of time-domain spectroscopy (TDS). The TDS uses femtosecond pulses of near-visible laser light to generate coherent THz waves with opto-electronical means. The resulting electromagnetic pulse is broadband and ranges from approximately 300 GHz to 3 THz (recently expanded to app. 10 THz). As it was mentioned before, there are natural sources of cosmic THz radiation: stars or any other "heated" celestial bodies with temperatures of 10 K and higher emanate THz, *i.e.* almost any heated body. The commercial sources (and detectors), on the other hand, were difficult to implement up until the last decade or so because of technological and fundamental difficulties.

Chapter 2

Basics of Electromagnetism

Terahertz (THz) waves are part of electromagnetic radiation (EMR) that takes the form of self-propagating waves in a vacuum or in matter. The EMR parts are: radio waves, microwaves, THz radiation, infrared radiation, visible light, ultraviolet radiation, X-rays and gamma rays. As EMR, THz has electric and magnetic field components, which oscillate in phase perpendicular to each other and the direction of energy propagation.

According to the Maxwell's theory of electromagnetism, a time-varying electric field generates a magnetic field and vice versa. Therefore, as an oscillating electric field generates an oscillating magnetic field, the magnetic field in turn generates an oscillating electric field, and so on. These oscillating fields together form an electromagnetic wave.

2.1. Nature of electromagnetism

Our physical universe is governed by four fundamental forces of nature:

Nuclear force that is strongest of the four, but its range is limited to submicroscopic systems, such as nuclei.

Weak-interaction force whose strength is only 10^{-14} that of the nuclear force. Its primary role is in interactions involving certain radioactive elementary particles.

Electromagnetic force exists between all charged particles. It is the dominant force in **microscopic** systems, such as atoms and molecules, and its strength is in the order of 10^{-2} that of the nuclear force.

Gravitational force is the weakest of all four forces, having a strength in the order of 10^{-41} that of the nuclear force. However, it is the dominant force in **macroscopic** systems, such as the solar system.

Even though the electromagnetic force operates on the atomic scale, its effects can be transmitted in the form of electromagnetic waves that can propagate through both free space and material media.

2.2. Electric fields

The electromagnetic force consists of an electrical force $\mathbf{F_e}$ and a magnetic force $\mathbf{F_m}$. The electrical force $\mathbf{F_e}$ is similar to the gravitational force, but with a major difference. The source of the gravitational force is mass, and the source of the electrical field is electrical charge, and whereas both types of fields vary inversely as the square of the distance from their respective sources, electric charge may have positive or negative polarity, whereas mass does not exhibit such a property.

We know from atomic physics that all matter contains a mixture of neutrons, positively charged protons, and negatively charged electrons, with the fundamental quantity of charge being that of a single electron, usually denoted by the letter "e". The unit by which electric charge is measured is the Coulomb (C), named after of the 18th-century French scientist Charles Augustin de Coulomb (1736–1806). The magnitude of e is

$$e = 1.6 \times 10^{-19} \, \text{C}. \tag{2.1}$$

The charge of a single electron is $q_e = -e$ and that of a photon is equal in magnitude but opposite in polarity: $q_p = e$. Coulomb's experiments demonstrated that:

(1) Two like charges repel one another, whereas two charges of opposite polarity attract.
(2) The force acts along the line joining the charges.
(3) The strength of this force is proportional to the product of the magnitudes of the two charges, and inversely proportional to the square of the distance between them.

These properties constitute what today is called **Coulomb's Law**, which can be expressed mathematically by the following equation:

$$\vec{F}_{e21} = \vec{R}_{12} \frac{q_1 q_2}{4\pi\varepsilon_0 R_{12}^2} \text{N} \quad \text{(in free space)}, \tag{2.2}$$

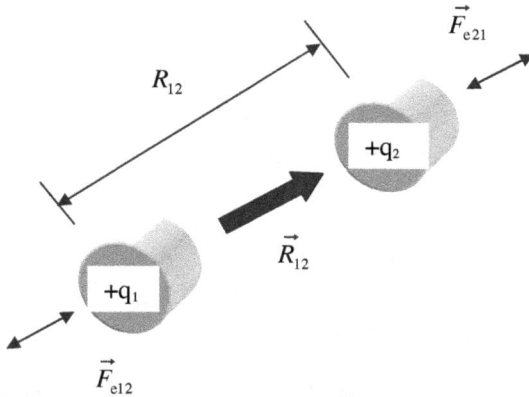

Fig. 2.1.   Electric forces on two positive point charges in free space.

where $\vec{F}_{e21}$ is the electrical force acting on charge q_2 due to charge q_1, R_{12} is the distance between the two charges, $\vec{R}_{12}$ is a unit vector pointing from charge q_1 to charge q_2 (Fig. 2.1) and ε_0 is a universal constant called the **electrical permittivity of free space** [$\varepsilon_0 = 8.854 \times 10^{-12}$ Farad per meter (F/m)].

The two charges are assumed to be in **free space** (vacuum) and isolated from all other charges. The force $\vec{F}_{e12} = -\vec{F}_{e21}$.

The existence of an **electric field intensity, E** due to any charge q is as follows:

$$\vec{E} = \vec{R}\frac{q}{4\pi\varepsilon_0 R^2} \text{ V/m} \quad \text{(free space)}, \tag{2.3}$$

where R is the distance between the charge and the observation point, and $\vec{R}$ is the unit vector pointing away from the charge. It is significant to point out that (2.3) is analogous to the Newton's law of gravity:

$$\vec{F}_{g21} = -\vec{R}_{12}\frac{G_{m_1}G_{m_2}}{R_{12}^2}\text{N}, \tag{2.4}$$

where $\vec{R}_{12}$ is a unit vector that points from mass m_1 to mass m_2, G is the universal gravitational force and F_{g21} is in Newtons (N). The negative sign reflects the fact that the gravitational force is attractive. This is one of many examples of similarity between the laws of mechanics and those of electromagnetism.

Figure 2.2 depicts the electric-field lines due to a positive charge. $\vec{E}$ is measured in volt per meter (V/m).

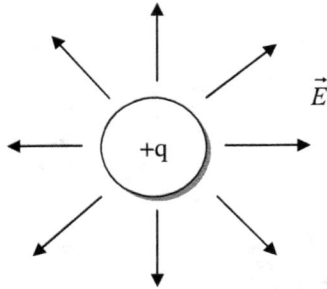

Fig. 2.2.   Electric field $\vec{E}$ due to a charge q.

Electric charge exhibits two important properties. The first is the **law of conservation of electric charge** that postulates that the net electric charge cannot be created or destroyed. If a volume contains $\mathbf{n_p}$ protons and $\mathbf{n_e}$ electrons, then the total charge is:

$$q = n_p e - n_e e = (n_p - n_e)e \text{ C.} \tag{2.5}$$

Even if some of the photons combine with an equal number of electrons to produce neutrons or other elementary particles, the net charge q remains constant. In matter, the quantum mechanical laws governing the behavior of the protons inside the atom's nucleus and the electrons outside it, do not allow them to combine.

The second important property of electric field is the **principle of linear superposition**, which states that *the total vector electric field at a point in space due to a system of point charges is equal to the vector sum of the electric fields at that point due to the individual charges.*

This seemingly simple concept will allow us to compute the electric field due to complex distribution of charge without having to be concerned with the forces acting on each individual charge due to the fields by all of the other charges.

The expression (2.3) describes the field induced by an electric charge in free space. Let us now consider what happens when we place a positive point charge in a material composed of atoms.

In the absence of the point charge, the material is electrically neutral, with each atom having a positively charged nucleus surrounded by a cloud of electrons of equal but opposite polarity. Hence, at any point in the material not occupied by an atom the electric field E is zero. Upon placing a point charge in the material, as shown in Fig. 2.3, the atoms experience forces that cause them to become ordered in a certain way.

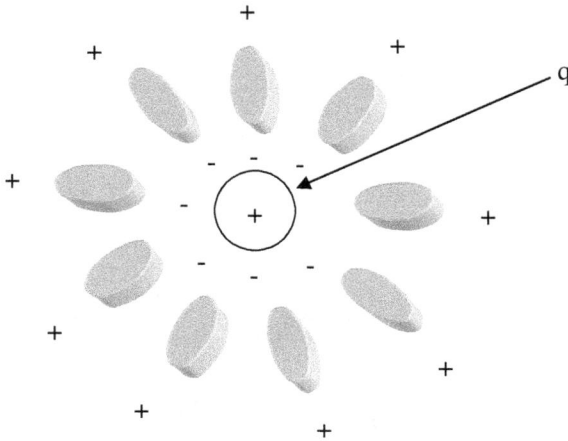

Fig. 2.3.   Polarization of the atoms of a dielectric material by a positive charge q.

The center of symmetry of the electron cloud is altered with respect to the nucleus, with one pole of the atom becoming more positively charged and the other pole becoming more negatively charged. Such a polarized atom is called an **electric dipole**, and the process of the charge distribution is called **polarization**. The degree of polarization depends on the distance between the atom and the isolated point charge, and the orientation of the dipole is such that the dipole axis connecting its two poles is directed toward the point charge as schematically illustrated in Fig. 2.3. The net result of this polarization process is that the electric dipoles of the atoms (or molecules) tend to counteract the field due to the point charge. Consequently, the electric field at any point in the material would be different from the field that would have been induced by the point charge in the absence of the material.

In order to extend Eq. (2.3) to be applicable in any medium, the permittivity of free space ε_0 with ε is introduced, where ε is now permittivity of the material in which the electrical field is measured and is, therefore, descriptive of that particular material. Thus,

$$\vec{E} = \vec{R}\frac{q}{4\pi\varepsilon R^2} \text{ V/m} \tag{2.6}$$

ε is usually expressed in the form:

$$\varepsilon = \varepsilon_r\varepsilon_0 \text{ F/m}, \tag{2.7}$$

Table 2.1.   Dielectric constants (relative permittivity, ε_r).

Material	Dielectric constant, ε_r	Material	Dielectric constant, ε_r
Vacuum	1	Dry soil	2.5–3.5
Air (at sea level)	1.0006	Plexiglas	3.4
Styrofoam	1.03	Glass	4.5–10
Teflon	2.1	Quartz	3.8–5
Wood (dry)	1.5–4	Bakelite	5
Polyethylene	2.25	Porcelain	5.4–6
Polystyrene	2.6	Mica	5.4–6
Paper	2–4	Ammonia	22
Rubber	2.2–4.1	Distilled water	81

where ε_r is a dimensionless quantity called the **relative permittivity** or **dielectric constant** (more often) of the material. For vacuum, $\varepsilon_r = 1$; for air near the surface, $\varepsilon_r = 1.0006$.

In addition to the electric field intensity $\vec{E}$, it is often convenient to use a related quantity called the **electric flux density, D**, given by

$$\vec{D} = \varepsilon\vec{E}\,\text{C/m}^2, \tag{2.8}$$

where $\vec{E}$ and $\vec{D}$ constitute important parts of the theory of electromagnetism.

The electric flux density is expressed in Coulomb per square meter (C/m^2).

2.3. Magnetic fields

The magnetic lines encircling a magnet (Fig. 2.4) are called **magnetic field lines** and represent the existence of a magnetic field called the **magnetic flux density, B**. A magnetic field not only exists around permanent magnets but can also be caused by electric current. The connection between electricity and magnetism was discovered by the Danish scientist Hans Oersted in 1819.

Shortly after Oersted's discovery, French scientists Jean Baptiste Biot and Felix Savart developed an expression that relates the magnetic field $\vec{B}$ at a point in space to the current I in the conductor. The derived relation is known as Biot–Savart law (2.9).

$$\vec{B} = \vec{\varphi}\frac{\mu_0 I}{2\pi r}\,\text{T}. \tag{2.9}$$

The formulation of the law states that the **magnetic flux density**, $\vec{B}$ is proportional to the current in a long wire and inversely proportional to the

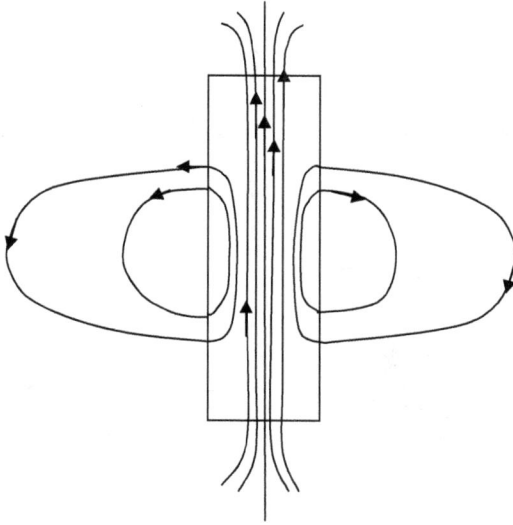

Fig. 2.4. Pattern of magnetic field lines around a bar magnet.

radial distance, r from the wire with current. $\vec{\varphi}$ is an azimuthal unit vector denoting the fact that the magnetic field direction is tangential to the circle surrounding the current (Fig. 2.5). The magnetic field is measured in Tesla (T), named after Nicola Tesla whose work on transformers made it possible to transport electricity over long wires without much loss. The quantity μ_0 is called the **magnetic permeability of free space** [$\mu_0 = 4\pi \times 10^{-7}\,\mathrm{H/m}$, where H/m is Henry per meter]. This unit is analogous to the electric permittivity ε_0. Both units are connected in the expression for the speed of light, c.

$$c = \frac{1}{\sqrt{\mu_0 \varepsilon_0}} = 3 \times 10^8\,\mathrm{m/s}. \tag{2.10}$$

The majority of natural materials are **nonmagnetic**, meaning that they exhibit a magnetic permeability $\mu = \mu_0$. For ferromagnetic materials, such as iron and nickel, μ can be much larger than μ_0. The magnetic permeability μ accounts for **magnetization** properties of a material. Similar to (2.7):

$$\mu = \mu_r \mu_0\,\mathrm{H/m}, \tag{2.11}$$

where μ_r is a dimensionless quantity called the **relative magnetic permeability** of the material. The values of some μ_r are given in Table 2.2.

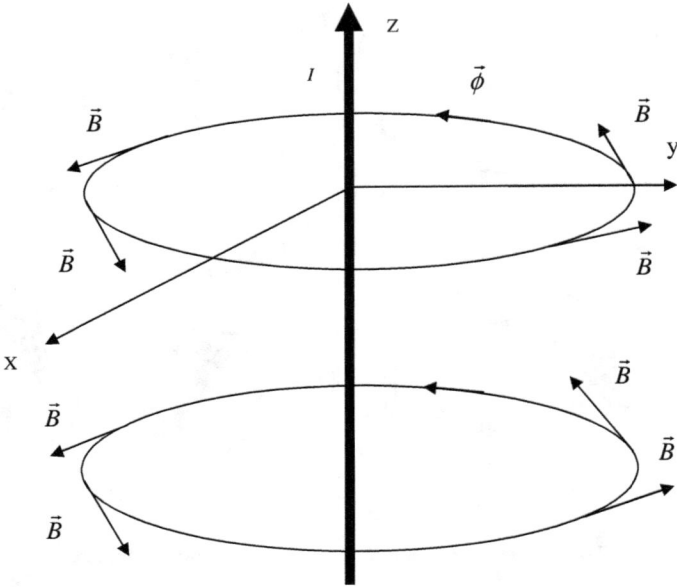

Fig. 2.5.   Magnetic field induced by a steady current flowing in the z-direction.

Table 2.2.   Relative permeability μ_r of common materials, $\mu_0 = 4\pi \times 10^{-7}$ H/m.

Material	Relative permeability, μ_r	Material	Relative permeability, μ_r
Diamagnetic		*Paramagnetic*	
Bismuth	0.99983	Air	1.000004
Gold	0.99996	Aluminum	1.00002
Mercury	0.99997	Tungsten	1.00008
Silver	0.99998	Titanium	1.0002
Copper	0.99999	Platinum	1.0003
Water	0.99999		
Ferromagnetic			
Cobalt	250		
Nickel	600		
Mild steel	2,000		
Iron (pure)	4,000–5,000		
Silicon iron	7,000		
Purified iron	~200,000		

Note: Except for ferromagnetic materials, μ_r for all dielectrics and conductors.

Analogous to the electric flux density, D (2.8), we have an expression for the **magnetic field intensity**, H that are related to each other through μ:

$$\vec{B} = \mu \vec{H}. \tag{2.12}$$

2.4. Static and dynamic fields

Since the electric field $\vec{E}$ is governed by the charge q and the magnetic field $\vec{H}$ is governed by $I = \frac{dq}{dt}$, and since q and $\frac{dq}{dt}$ are independent variables, the induced electric and magnetic fields are independent of one another as long as I remains constant. To demonstrate the validity of this statement, consider, for example, a section of a beam of charged particles moving at a constant velocity. The moving charge constitutes a DC current. The electric field due to that section of the beam is determined by the total charge q contained in that section. The magnetic field does not depend on q, but rather on the rate of charge (that is carried by the current) flowing in that section. Few charges moving fast can constitute the same current as many charges moving slowly. In these two cases the induced magnetic field will be the same because the current I is the same, but the induced electric field will be quite different because the amount of charge is not the same. **Electrostatics and magnetostatics**, corresponding to stationary charges and steady currents, respectively, are special cases of electromagnetics. They represent two independent branches, so characterized because the induced electric and magnetic fields are uncoupled from each other. **Dynamics**, the third and more general branch of electromagnetics, involves time-varying fields induced by time-varying sources, that is, currents and charge densities. If the current associated with the beam of moving charged

Table 2.3. Three branches of electromagnetic.

Branch	Condition	Field quantities (Units)
Electrostatics	Stationary charges $(\delta q/\delta t = 0)$	Electric field intensity $\vec{E}$(V/m) Electric flux density $\vec{D}$(C/m^2) $\vec{D} = \varepsilon\vec{E}$
Magnetostatics	Steady currents $(\delta I/\delta t = 0)$	Magnetic flux density $\vec{B}$(T) Magnetic field density $\vec{H}$(A/m) $\vec{B} = \mu\vec{H}$
Dynamics *(Time-varying fields)*	Time-varying currents $(\delta I/\delta t0 \neq 0)$	$\vec{E}, \vec{D}, \vec{B}, \vec{H}$ $(\vec{E}, \vec{D})$ coupled to $(\vec{B}, \vec{H})$

Table 2.4.   Constitutive parameters of materials.

Parameters	Unit	Free space value
Electrical permittivity, ε	F/m	$\varepsilon_0 = 8.854 \times 10^{-12}$ (F/m) $\approx \dfrac{1}{36} \times 10^{-9}$ (F/m)
Magnetic permeability, μ	H/m	$\mu_0 = 4\pi \times 10^{-7}$ (H/m)
Conductivity, σ	S/m	0

particles varies with time, then the amount of charge present in a given section of the beam also varies with time, and vice versa.

A time-varying electric field will generate a time-varying magnetic field and vice versa.

2.5. Maxwell's equations

Electromagnetism is based on a set of four fundamental relations known as Maxwell's equations:

$$\nabla \cdot \vec{D} = \rho_v \quad \text{(Gauss's Law)}, \tag{2.13}$$

$$\nabla \times \vec{E} = -\frac{\delta \vec{B}}{\delta t} \quad \text{(Faraday's Law)}, \tag{2.14}$$

$$\nabla \cdot \vec{B} = 0 \quad \text{(Gauss's Law) (magnetic charge does not exist)} \tag{2.15}$$

$$\nabla \times \vec{H} = \vec{J} + \frac{\delta \vec{D}}{\delta t} \quad \text{(Ampere's Law)}, \tag{2.16}$$

where $\vec{E}, \vec{D}$ are electric field quantities interrelated by $\vec{D} = \varepsilon \vec{E}$ with ε being the electrical permittivity of the material; $\vec{B}, \vec{H}$ are magnetic field quantities interrelated by $\vec{B} = \mu \vec{H}$, with μ being the magnetic permeability of the material; ρ_v is the electric charge per unit volume; and $\vec{J}$ is the current density per unit area. The field quantities $\vec{E}, \vec{D}, \vec{B}$ and $\vec{H}$ were introduced before. These equations hold for any material, including free space (vacuum), and at any spatial location (x, y, z). In general, all the quantities in Maxwell's equations may be a function of time t. In 1873, James Clerk Maxwell established the first unified theory of electricity and magnetism. The equations, which he deduced from experimental observations reported by Gauss, Ampere, Faraday, and others, not only encapsulate the connection between the electric field and electric charge, as well as between the magnetic field and electric current but also define the bilateral coupling between the electric and magnetic field quantities. Combined with auxiliary relations, Maxwell's equations form the fundamental tenets of electromagnetic theory.

In the static case, none of the quantities appearing in Maxwell's equations are a function of time (*i.e.* $\frac{\delta}{\delta t} = 0$). This takes place when all charges are permanently fixed in space or move at a constant speed. In this case, ρ_v and J are constant in time. Under these circumstances, the time derivatives of $\vec{B}$ and $\vec{D}$ in (2.14) and (2.16) are zero, and Maxwell's equations reduce to:

Electrostatics	$\nabla \cdot \vec{D} = \rho_v,$	(2.17)
	$\nabla \times \vec{E} = 0,$	(2.18)
Magnetostatics	$\nabla \cdot \vec{B} = 0,$	(2.19)
	$\nabla \times \vec{H} = \vec{J}.$	(2.20)

Maxwell's four equations separate into two pairs, with the first pair involving only the electric quantities $\vec{E}$ and $\vec{D}$ and the second pair involving only the magnetic field quantities $\vec{B}$ and $\vec{H}$. This allows studying electricity and magnetism as two distinct and separate phenomena, as long as the spatial distributions of charge and current flow remain constant in time.

We study electrostatics not only as a prelude to the study of time-varying fields, but also because it is an important field of study in its own right. Many electronic devices and systems are based on the principles of electrostatics. They include X-ray machines, oscilloscopes, ink-jet electrostatic printers, liquid crystal displays, copying machines, capacitance keyboards, and many solid-state control devices. Electrostatics is also used in the design of medical diagnostic sensors, such as the electrocardiogram, and the electroencephalogram (brain activity recording), as well as in numerous industrial applications.

2.6. Coulomb's law

One of the major goals of this chapter is to develop expressions relating the **electric field**, $\vec{E}$ and the associated **electric flux density**, $\vec{D}$ to any specified distribution of charge. Our discussions, however, will be limited to electrostatic fields induced by static charge distributions. Coulomb's law, which was introduced for electrical charges in air and later generalized to material media, states that:

(1) An isolated charge q induces an electric field $\vec{E}$ at every point in space, and at any specific point P, $\vec{E}$ is given by:

$$\vec{E} = \vec{R}\frac{q}{4\pi\varepsilon R^2},$$
(2.21)

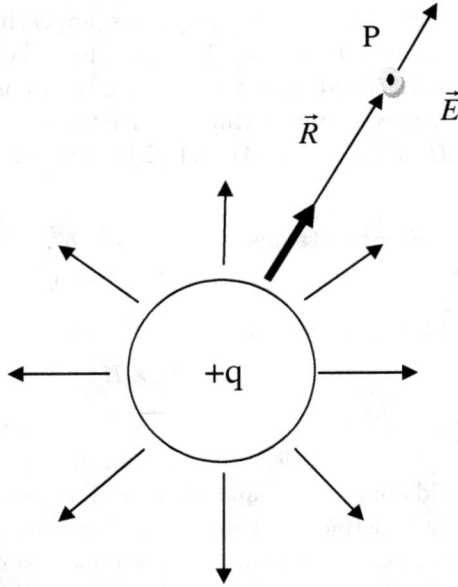

Fig. 2.6.   Electric-field lines for a single charge.

where $\vec{R}$ is a unit vector pointing from q to P (Fig. 2.6). R is the distance between them, and ε is the electrical permittivity of the medium containing the observation point P, and q.

(2) In the presence of an electric field $\vec{E}$ at a given point in space, which may be due to a single charge or a distribution of many charges, the force acting on a test charge q', when the charge is placed at that point, is given by:

$$\vec{F} = q'\vec{E}\,\text{N},\qquad\qquad(2.22)$$

where $\vec{F}$ is measured in Newtons (N) and q' in Coulombs (C), the unit of $\vec{E}$ is in N/C or in V/m^2.

For a material with electrical permittivity ε, the electrical field quantities $\vec{E}$ and $\vec{D}$ are related by:

$$\vec{D} = \varepsilon\vec{E},\qquad\qquad(2.23)$$

where

$$\varepsilon = \varepsilon_{\mathrm{r}}\varepsilon_0;\quad \varepsilon_0 = 8.85 \times 10^{-12} \approx \frac{1}{36\pi} \times 10^{-9}\,\text{F/m}$$

is the electrical permittivity of free space, and $\varepsilon = \varepsilon_r \varepsilon_0$ is called the relative permittivity (or dielectric constant) of the material.

For most materials and under most conditions, ε of the material has a constant value independent of both the magnitude and direction of $\vec{E}$. Materials do not usually exhibit nonlinear permittivity behavior except when the amplitude of $\vec{E}$ is very high (at levels approaching the dielectric breakdown conditions), and anisotropy is peculiar only to certain materials with particular crystalline structures. Hence, except for materials under these very special circumstances, the quantities $\vec{D}$ and $\vec{E}$ are effectively redundant; for a material with known ε, knowledge of either $\vec{D}$ or $\vec{E}$ is sufficient to specify the other in that material.

2.7. Gauss's law

Now return to Eq. (2.13):

$$\nabla \cdot \vec{D} = \rho_v \quad \text{(Gauss's law)}, \qquad (2.24)$$

which is called the differential form of the Gauss's Law. "Differential" means that the divergence operation involves spatial derivatives. Equation (2.24) can be converted and expressed in the integral form. While solving electromagnetic problems, we often convert back and forth between the differential and integral forms of equations, depending on which form happens to be more applicable or convenient to use in a certain situation. To convert (2.24) into the integral form, we multiply both sides by dv and take the volume integral over an arbitrary volume v. Thus,

$$\int_v \nabla \cdot \vec{D} dv = \int_v \rho_v dv = Q, \qquad (2.25)$$

where Q is the total charge enclosed in v. The divergence theorem states that the volume integral of the divergence of any vector over a volume v is equal to the total outward flux of that vector through the surface s inclosing v. Thus, for the vector $\vec{D}$,

$$\int_v \nabla \cdot \vec{D} dv = \oint_s \vec{D} \cdot d\vec{s}. \qquad (2.26)$$

Comparison of (2.25) with (2.26) gives

$$\oint_s \vec{D} \cdot d\vec{s} = Q \quad \text{(Gauss's Law)}. \qquad (2.27)$$

The integral form of Gauss's Law is illustrated diagrammatically in Fig. 2.7.

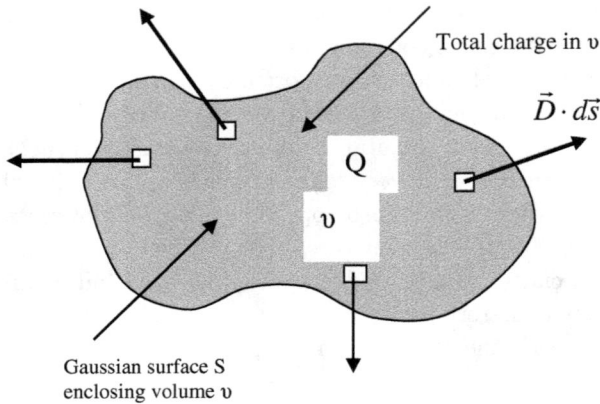

Fig. 2.7.   Gauss's Law: The outward flux of $\vec{D}$ through a surface is proportional to the enclosed charge Q.

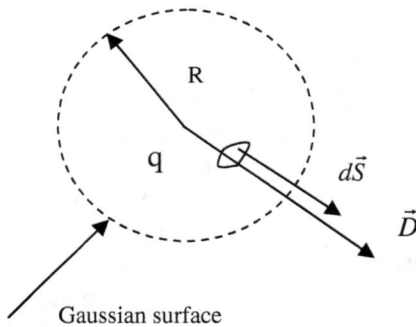

Fig. 2.8.   Electric field $\vec{D}$ due to a point charge q.

For each differential surface element of $d\vec{S}$, $\vec{D} \cdot d\vec{s}$ is the electric field flux flowing outwardly through $d\vec{S}$, and the total flux through the surface S is equal to the enclosed charge Q. The surface S is called a Gaussian surface.

When the dimensions of a very small volume Δv containing a total charge q are much smaller than the distance from Δv to the point at which the electric flux density $\vec{D}$ is to be evaluated, then q may be regarded as a **point charge**. The integral form of Gauss's Law can be applied to determine $\vec{D}$ due to a single isolated charge q by constructing a closed, spherical, Gaussian surface S of an arbitrary radius R centered at q, as shown in Fig. 2.8.

From symmetry considerations, assuming that q is positive, the direction of $\vec{D}$ must be radially outward along the unit vector $\hat{R}$, and D_R, which is the magnitude of $\vec{D}$, must be the same at all points on the Gaussian surface S. Thus, at any point on the surface, defined by position vector $\vec{R}$,

$$\vec{D}(\vec{R}) = \vec{R} D_R \tag{2.28}$$

and $d\vec{S} = \vec{R} ds$. Applying Gauss's Law gives:

$$\oint_s \vec{D} \cdot d\vec{S} = \oint_s \vec{R} D_R \cdot \vec{R} ds = \oint_s D_R ds = D_R(4\pi R^2) = q. \tag{2.29}$$

Solving for D_R and then inserting the result in (2.28) gives the following expression for the electric field $\vec{E}$ induced by an isolated point charge in a medium with permittivity ε:

$$\vec{E}(\vec{R}) = \frac{\vec{D}(\vec{R})}{\varepsilon} = \vec{R} \frac{q}{4\pi\varepsilon R^2} \text{ V/m}. \tag{2.30}$$

This is identical with Eq. (2.21) obtained from Coulomb's Law. For this simple case of an isolated charge, it does not matter much whether Coulomb's Law or Gauss's Law is used to obtain the expression for $\vec{E}$. However, it is more convenient to use Gauss's Law for any specified distribution of charge although its usefulness is limited to symmetrical charge distributions.

Gauss's Law as given by Eq. (2.24), provides a convenient method for determining the electrostatic flux density $\vec{D}$ when the charge distribution possesses symmetry properties that allow us to make valid assumptions about the variations in the magnitude and direction of $\vec{D}$ as a function of spatial location. Since at every point on the surface, the direction of $d\vec{S}$ is the outward normal to the surface, only the normal component of $\vec{D}$ at the surface contributes to the integral in Eq. (2.27). In order to successfully apply Gauss's Law, the surface S should be chosen such that from symmetry considerations, the magnitude of $\vec{D}$ is constant and its direction is normal or tangential at every point of each sub-surface of S (the surface of a cube, for example, has six sub-surfaces).

2.8. Poisson's equation

With $\vec{D} = \varepsilon\vec{E}$, the differential form of Gauss's Law given by Eq. (2.24) may be written as:

$$\nabla \cdot \vec{E} = \frac{\rho_v}{\varepsilon}. \tag{2.31}$$

Considering electrical potential as a function of electric field, we can write:

$$dV = \frac{dW}{q} = -\vec{E} \cdot \vec{dl}, \text{ J/C} \quad \text{or} \quad V, \tag{2.32}$$

where W is the work in joule to move an object over a vector differential distance $\vec{dl}$. The relationship between V and $\vec{E}$ for any charge distribution is given by:

$$\vec{E} = -\nabla V. \tag{2.33}$$

Inserting (2.33) in (2.31) gives:

$$\nabla \cdot (\nabla V) = -\frac{\rho_v}{\varepsilon}. \tag{2.34}$$

In view of the definition for the Laplacian of a scalar function V is given as:

$$\nabla^2 V = \nabla \cdot (\nabla V) = \frac{\delta^2 V}{\delta x^2} + \frac{\delta^2 V}{\delta y^2} + \frac{\delta^2 V}{\delta z^2}. \tag{2.35}$$

Equation (2.34) can be written in the abbreviated form:

$$\nabla^2 V = -\frac{\rho_v}{\varepsilon} \quad \text{(Poisson's equation).} \tag{2.36}$$

2.9. Electrical properties of materials

The electromagnetic constitutive parameters of a material medium are its electrical permittivity ε, magnetic permeability μ, and conductivity σ. A material is called *homogeneous* if its constitutive parameters do not vary from point to point, and it is *isotropic* if its constitutive parameters are independent of direction. Most materials exhibit isotropic properties but some crystals do not. In this section of the book, all materials are assumed to be homogeneous and isotropic.

The conductivity of a material is a measure of how easily electrons can travel through the material under the influence of an external electric field. Materials are classified as *conductors* (metals) or *dielectrics* (insulators) according to the magnitudes of their conductivities. A conductor has a large number of loosely attached electrons in the outermost shells of the atoms. In the absence of an external electric field, these free electrons move randomly producing zero average current through the conductor. Upon applying an external electric field, however, the electrons migrate from one atom to the next along a direction corresponding to that of the external field. Their movement, which is characterized by an average velocity, *called the electron drift velocity,* $\boldsymbol{u_e}$ gives rise to a *conduction current.*

In a dielectric, the electrons are tightly held to the atoms, so strongly that it is very difficult to detach them even under the influence of an electric field. Consequently, no current flows through the material. A *perfect dielectric* is a material with $\sigma = 0$ and, in contrast, a *perfect conductor* is a material with $\sigma = \infty$.

The conductivity σ of most metals is in the range from 10^6 to 10^7 S/m. Compared with 10^{-10} to 10^{-17} S/m for good insulators. Materials whose conductivities fall between those of conductors and insulators are called *semiconductors*. For example, the conductivity of pure germanium is 2.2 S/m.

The conductivity of a material depends on several factors, including temperature and the presence of impurities. In general, σ of metals increases with decreasing temperature, and at very low temperatures (close to absolute zero) some conductors become *superconductors* because their conductivity become almost infinity.

2.10. Conductors

The drift velocity $\vec{u}_e$ of electrons in a conducting material is related to the externally applied electric field $\vec{E}$ through

$$\vec{u}_e = -\mu_e \vec{E} \, \text{m/s}, \tag{2.37}$$

where μ_e is a material property called the *electron mobility* with units of $(m^2/V \cdot s)$. In a semiconductor, current flow is due to the movement of both electrons and holes, and since holes are positive-charge carriers, the *hole drift velocity* $\vec{u}_h$ is in the same direction as $\vec{E}$,

$$\vec{u}_h = \mu_h \vec{E} \, \text{m/s}, \tag{2.38}$$

where μ_h is the *hole mobility*. The mobility accounts for the *effective mass* of a charged particle and the average distance over which the applied electric field can accelerate it before it is stopped by colliding with an atom. In solid-state physics, a particle's *effective mass* is the mass that it would have in the semiconductor model of transport in a crystal. It may be demonstrated that electrons and holes in a crystal respond to electric and magnetic fields almost as if they were particles with a mass dependence in the direction of motion. The effective mass is a coefficient by which the electron mass is multiplied. It varies from 0.01 to 10 but can be as high as 1,000. Further, the current density can be expressed as:

$$\vec{J} = \rho_v \vec{u} \text{ A/m}^2, \tag{2.39}$$

where ρ_v = volume charge density; $\vec{u}$ = mean velocity of charged particles.

In the present case, the current density consists of a component $\vec{J}_e$ due to the electrons and a component $\vec{J}_h$ due to the holes. Thus, the total *conduction current density is*

$$\vec{J} = \vec{J}_e + \vec{J}_h = \rho_{ve}\vec{u}_e + \rho_{vh}\vec{u}_h \text{ A/m}^2. \tag{2.40}$$

Combining (2.37) and (2.38) gives:

$$\vec{J} = (-\rho_{ve}\mu_e + \rho_{vh}\mu_h)\vec{E}, \tag{2.41}$$

where $\rho_{ve} = -N_e e$ and $\rho_{vh} = N_h e$, with N_e and N_h being the number of free electrons and the number of free holes per unit volume, and $e = 1.6 \times 10^{-19} C$ is the absolute charge of a single hole or electron. The quantity inside the parentheses in Eq. (2.41) is defined as the *conductivity* of the material, σ. Thus,

$$\sigma = -\rho_{ve}\mu_e + \rho_{vh}\mu_h = (N_e\mu_e + N_h\mu_h)e \quad \text{S/m (for a semiconductor)} \tag{2.42}$$

and its unit is Siemens per meter (S/m). For a good conductor, $N_h\mu_h \ll N_e\mu_e$, and Eq. (2.42) reduces to:

$$\sigma = -\rho_{ve}\mu_e = N_e\mu_e e \quad \text{S/m (conductor).} \tag{2.43}$$

In either case, Eq. (2.41) becomes:

$$\vec{J} = \sigma\vec{E} \text{ A/m} \quad (Ohm's Law) \tag{2.44}$$

and it is called the point form of *Ohm's Law.*

For perfect dielectric: $\vec{J} = 0$.

For perfect conductor: $\vec{E} = 0$.

Since σ is on the order of 10^6 S/m for most metals, such as silver, copper, gold, and aluminum, it is common to set $\vec{E} = 0$ in metal conductors.

A perfect conductor is an equipotential medium, which means that the electric potential is the same at every point in the conductor. This property follows from the fact that V_{21}, the voltage difference between two points

(P_2 and P_1) in the conductor, is by definition equal to the line integral of $\vec{E}$ between two points, as indicated by:

$$V_{21} = V_2 - V_1 = -\int_{P_1}^{P_2} \vec{E} \cdot \vec{dl}, \qquad (2.45)$$

where $\vec{dl}$ is a vector differential distance.

Since $\vec{E} = 0$ everywhere in the perfect conductor, the voltage difference $V_{21} = 0$. The fact that the conductor is an equipotential medium, however, does not necessarily imply that the potential difference between two conductors is zero. Each conductor is an equipotential medium, but the presence of different distributions of charges on their surfaces can generate a potential difference between them.

2.11. Resistance

In order to demonstrate the use of the point form of Ohm's Law, we shall use it to derive an expression for the resistance R of a conductor of length $l = x_2 - x_1$. A voltage V applied across the conductor terminals establishes an electric field $\vec{E} = \vec{x} E_x$; the direction of $\vec{E}$ is from the point of the higher potential (point 1 in Fig. 2.9) to the point of the lower potential (point 2). The relation between V and E_x is obtained by applying Eq. (2.45). Thus,

$$V = V_1 - V_2 = -\int_{x_2}^{x_1} \vec{E} \cdot \vec{dl} = -\int_{x_2}^{x_1} \vec{x} E_x \cdot \vec{x} dl = E_x l, V \qquad (2.46)$$

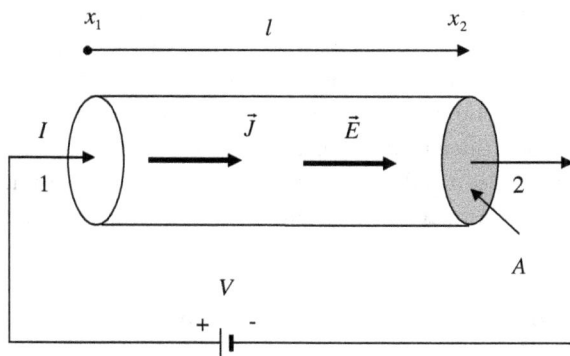

Fig. 2.9. Linear resistor with cross-section A and length l connected to a DC voltage source V.

Using Eq. (2.44), the current flowing through the cross-section A at x_2 is:

$$I = \int_A \vec{J} \cdot d\vec{s} = \int_A \sigma \vec{E} \cdot d\vec{s} = \sigma E_x A, \, A. \tag{2.47}$$

From $R = V/I$, the ratio of Eq. (2.46) to Eq. (2.47) gives:

$$R = \frac{l}{\sigma A}, \, \Omega. \tag{2.48}$$

We now generalize our result for R to any resistor of arbitrary shape by noting that the voltage V across the resistor is equal to this line integral of $\vec{E}$ over a path l between two specified points and the current I is equal to the flux of $\vec{J}$ through the surface S of the resistor. Thus,

$$R = \frac{-\int_l \vec{E} \cdot d\vec{l}}{\int_l \vec{J} \cdot d\vec{s}} = \frac{-\int_l \vec{E} \cdot d\vec{l}}{\int_S \sigma \vec{E} \cdot d\vec{s}} = \frac{V}{I}. \tag{2.49}$$

The reciprocal of R is called the *conductance* G, and the unit of G is (Ω^{-1}) or Siemens (S). For the linear resistor:

$$G = \frac{1}{R} = \frac{\sigma A}{l} \text{S}. \tag{2.50}$$

2.12. Dielectrics

As we discussed previously, the fundamental difference between a conductor and a dielectric is that a conductor holds (free) electrons that can migrate through the crystalline structure of the material, whereas the electrons in the outermost shells of a dielectric are strongly bound to the atom. In the absence of an electric field, the electrons in any material form a symmetrical cloud around the nucleus with the center of the cloud being at the same location as the center of the nucleus, as shown in Fig. 2.10(a).

The electric field generated by the positively charged nucleus attracts and holds the electron cloud around it, and the mutual repulsion of the electron clouds of adjacent atoms gives matter its form. When a conductor is subjected to an externally applied electric field, the most loosely bound electrons in each atom can easily jump from one atom to the next, thereby setting up an electric current.

In a dielectric, however, an external applied electric field $\vec{E}_{\text{ext}}$ cannot effect mass migration of charges since none are able to move freely, but it can *polarize* the atoms or molecules in the material by distorting the center of the cloud and the location of the nucleus. The polarization process is illustrated in Fig. 2.10(b). The polarized atom or molecule maybe

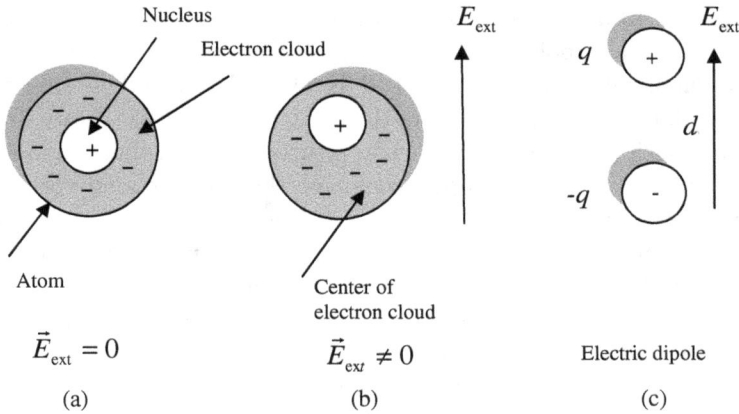

Fig. 2.10.   Position of the center of the nucleus and the electron cloud: (a) in the absence of an external electric field, $\vec{E}_{ext}$; (b) separation of the nucleus and the electronic cloud with $\vec{E}_{ext}$ applied; and (c) electric dipole with a separation, d.

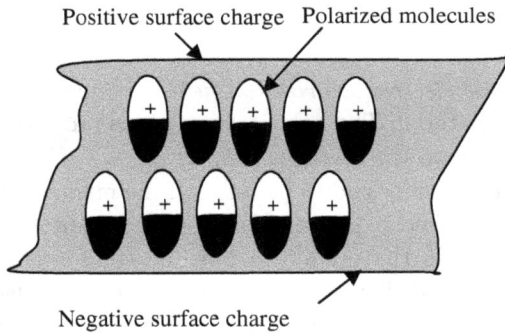

Fig. 2.11.   Dielectric medium polarized by an external electric field $\vec{E}_{ext}$ and a negative density on the lower surface.

represented by an electric dipole consisting of a charge $+q$ at the center of the nucleus and charge $-q$ at the center of the electron cloud (Fig. 2.10(c)). Each such dipole sets up a small electric field, pointing from the positively charged nucleus to the center of the equally but negatively charged electron cloud. This *induced* electric field, called a *polarization* field, is weaker than and opposite in direction to $\vec{E}_{ext}$. Consequently, the net electric field present in the dielectric material is smaller than $\vec{E}_{ext}$. At the microscopic level, each dipole exhibits a dipole moment. Within the dielectric material, the dipoles align themselves in a linear system, as shown in Fig. 2.11.

Figure 2.10 depicts *nonpolar materials* in which the molecules do not have permanent dipole moments. Nonpolar molecules become polarized only when an external electric field is applied, and when the field is terminated, the molecules return to their original unpolarized state. In some materials, such as water, the molecular structure is such that the molecules possess built-in permanent dipole moments that are randomly oriented in the absence of an applied electric field. Materials composed of permanent dipoles are called *polar materials*. Owing to their random orientations, the dipoles of polar materials produce no net dipole moment macroscopically (at the macroscopic) scale, each point in the material represents a small volume containing thousand of molecules). Under the influence of an applied field, an arrangement somewhat similar to that shown in Fig. 2.11 takes place for nonpolar and polar materials.

Whereas $\vec{D}$ and $\vec{E}$ are related by ε_0 in free space, the presence of these microscopic dipoles in a dielectric material alters that relationship in that material to:

$$\vec{D} = \varepsilon_0 \vec{E} + \vec{P}, \tag{2.51}$$

where $\vec{P}$, called the *electric polarization field,* accounts for the polarization properties of the material. The polarization field is produced by the electric field $\vec{E}$ and depends on the material properties.

A dielectric medium is said to be *linear* if the magnitude of the induced polarization field is directly proportional to the magnitude of the polarization field and $\vec{E}$ are in the same direction. In some crystals, the periodic structure of the material allows more polarization to take place along certain directions, such as the crystal axes than along others. In such *anisotropic* dielectrics, $\vec{E}$ and $\vec{D}$ may have different directions.

A medium is called *homogeneous* if its constitutive parameters $(\varepsilon, \mu, \text{ and } \sigma)$ are constant throughout the medium. Our present treatment will be limited to media that are linear, isotropic, and homogeneous. For such media, the polarization field is directly proportional to $\vec{E}$ and is expressed by the relationship:

$$\vec{P} = \varepsilon_0 \chi_e \vec{E}, \tag{2.52}$$

where χ_e is called the electric susceptibility of the material. Inserting (2.52) in (2.51), we obtain:

$$\vec{D} = \varepsilon_0 \vec{E} + \varepsilon_0 \chi_e \vec{E} = \varepsilon_0 (1 + \chi_e) \vec{E} = \varepsilon \vec{E}, \tag{2.53}$$

Table 2.5. Relative permittivity (dielectric constant) and
dielectric strength of common materials.

Material	Relative permittivity, ε_r	Dielectric strength, E_{ds}, (MV/m)
Air (at sea level)	1.0006	3
Petroleum oil	2.1	12
Polystyrene	2.6	20
Glass	4.5–10	25–40
Quartz	3.8–5	30
Bakelite	5	20
Mica	5.4–6	200

which defines the permittivity ε of the material as:

$$\varepsilon = \varepsilon_0(1 + \chi_e). \tag{2.54}$$

As it was mentioned earlier, it is often convenient to characterize the permittivity of a material relative to that of free space, ε_0 this is taken into account by the relative permittivity $\varepsilon_r = \varepsilon/\varepsilon_0$. Values of ε_r are listed in Table 2.5.

The value of ε_0 is $\varepsilon_0 = 8.854 \times 10^{-12}\,\mathrm{F/m}$.

Therefore, we obtain ε by using the following formula:

$$\varepsilon = \varepsilon_r \varepsilon_0.$$

In free space $\varepsilon_r = 1$, and for most conductors $\varepsilon_r \approx 1$. The dielectric constant is approximately 1.0006 at sea level, and it decreases toward unity with increasing altitude. In a number of electromagnetic applications, electromagnetic waves propagation through the atmosphere depends on absorption effects (*e.g.* for THz radiation). The absorption increases from $\sim$10 to $\sim$10^5 dB/km for the range 1–10 THz.[37]

The dielectric polarization model presented this far has placed no restriction on the upper end of the strength of the applied electric field $\vec{E}$. In reality, if $\vec{E}$ exceeds a certain critical value, known as the *dielectric strength* of the material, it will free the electrons completely from the molecules and cause them to accelerate through the material that can sustain permanent damage due to electron collision with the molecular structure. This abrupt change in behavior is called a *dielectric breakdown*. The dielectric strength E_{ds} is the highest magnitude of $\vec{E}$ that the material can sustain without breakdown. Dielectric breakdown can occur in gas, liquid, and solid dielectrics. The associated field strength depends on the material composition, as well as other factors such temperature and humidity.

2.13. Magnetostatics

This chapter on Magnetostatics provides an analogous treatment to that for electrostatics. Stationary charges produce static electric fields, and steady currents (not varying with time) produce magnetic fields. For $\delta/\delta t = 0$, the magnetic fields in a medium with magnetic permeability μ are governed by the second pair of Maxwell's equations, those given by Eqs. (2.19) and (2.20):

$$\nabla \cdot \vec{B} = 0, \tag{2.55}$$

$$\nabla \times \vec{H} = \vec{J}, \tag{2.56}$$

where $\vec{J}$ is the current density. The magnetic flux density $\vec{B}$ and the magnetic field intensity $\vec{H}$ are related by:

$$\vec{B} = \mu\vec{H}. \tag{2.57}$$

When we examined electric fields in a dielectric medium in the previous chapter, we noted that the relation $\vec{D} = \varepsilon\vec{E}$ is valid only when the medium is linear and isotropic. These properties, which are true for most materials, allow us to treat the permittivity ε as a constant scalar quantity, independent of both the magnitude and direction of $\vec{E}$. A similar statement applies to the relation given by Eq. (2.57). With the exception of ferromagnetic materials, for which the relationship between $\vec{B}$ and $\vec{H}$ is nonlinear, most materials are characterized by constant magnetic permeabilities. Furthermore, $\mu = \mu_0$ holds for most dielectric materials excluding ferromagnetic materials.

The parallelism between magnetostatic quantities and their electrostatic counterparts is shown in Table 2.6.

The electric field $\vec{E}$ at a point in space has been defined as the electric force $\vec{F}_e$ per unit charge acting on a test charge when placed at that point. We now define the *magnetic flux density* $\vec{B}$ at a point in space in terms of the *magnetic force* $\vec{F}_m$ that is exerted on a charged particle moving with a velocity $\vec{u}$ if it passes through that point. Based on experiments conducted to determine the motion of charged particles moving in magnetic fields, it was established that the magnetic force $\vec{F}_m$ acting on a particle of charge q can be expressed in the form:

$$\vec{F}_m = q\vec{u} \times \vec{B}, N \tag{2.58}$$

Accordingly, the strength of $\vec{B}$ is measured in Newtons/(C·m/s), which also is called Tesla (T) in SI units. For a positively charged particle, the

Table 2.6. Attributes of electrostatics and magnetostatics.

	Electrostatics	Magnetostatics
Sources	Stationary charges	Steady currents
Fields	$\vec{E}$ and $\vec{D}$	$\vec{H}$ and $\vec{B}$
Constitutive parameters	ε and σ	μ
Governing equations		
* Differential form	$\nabla \cdot \vec{D} = \rho_v$	$\nabla \cdot \vec{B} = 0$
	$\nabla \times \vec{E} = 0$	$\nabla \times \vec{H} = \vec{J}$
*Integral form	$\oint_s \vec{D} \cdot d\vec{s} = Q$	$\oint_s \vec{B} \cdot d\vec{s} = 0$
	$\oint_c \vec{E} \cdot d\vec{l} = 0$	$\oint_c \vec{H} \cdot d\vec{l} = I$
Potential	Scalar V with $\vec{E} = -\nabla V$	Vector $\vec{A}$ with $\vec{B} = \nabla \times \vec{A}$
Energy density	$\omega_e = \frac{1}{2}\varepsilon E^2$	$\omega_m = \frac{1}{2}\mu H^2$
Force on charge q	$\vec{F}_e = q\vec{E}$	$\vec{F}_m = q\vec{u} \times \vec{B}$
Circuit elements	C and R	L

direction of $\vec{F}_m$ is in the direction of the cross product $\vec{u} \times \vec{B}$ which is perpendicular to the plane containing $\vec{u}$ and $\vec{B}$ and governed by the right-hand rule. If q is negative, the direction of $\vec{F}_m$ is reversed, as illustrated in Fig. 2.12. The magnitude of $\vec{F}_m$ is given by:

$$F_m = quB \sin\theta, \tag{2.59}$$

where θ is the angle between $\vec{u}$ and $\vec{B}$. We note that F_m is maximum when $\vec{u}$ is perpendicular to $\vec{B}$ ($\theta = 90°$), and it is zero when $\vec{u}$ is parallel to $\vec{B}$ ($\theta = 0$ or $180°$).

If a charged particle is in the presence of both an electric field $\vec{E}$ and a magnetic field $\vec{B}$, then the total *electromagnetic force* acting on it is:

$$\vec{F} = \vec{F}_e + \vec{F}_m = q\vec{E} + q\vec{u} \times \vec{B} = q(\vec{E} + \vec{u} \times \vec{B}). \tag{2.60}$$

The force expressed by Eq. (2.60) is known as the *Lorentz force*. Electric and magnetic forces exhibited a number of important differences:

(1) Whereas the electric force is always in the direction of the electric field, the magnetic force is always perpendicular to the magnetic field.
(2) Whereas the electric force acts on a charged particle whether or not it is moving, the magnetic force acts on it only if it is in motion.

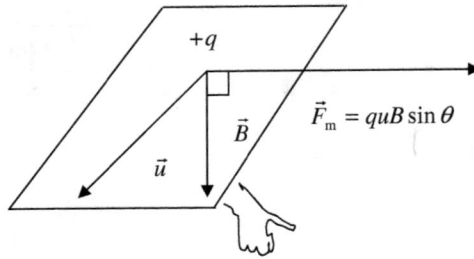

(a)

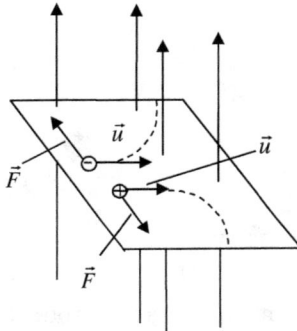

(b)

Fig. 2.12.   Direction of the magnetic force exerted on a charged particle moving in a magnetic field is (a) perpendicular to both $\vec{B}$ and $\vec{u}$; (b) depends on the charge polarity.

(3) Whereas the electric force expends energy in displacing a charged particle, the magnetic force does not work when a particle is displaced.

Our last statement requires further elaboration. Since the magnetic force $\vec{F}_m$ is always perpendicular to $\vec{u}$, $\vec{F} \cdot \vec{u} = 0$. Hence, the work performed when a particle is displaced by a differential distance $\vec{dl} = \vec{u}dt$ is:

$$dW = \vec{F}_m \cdot \vec{dl} = (\vec{F}_m \cdot \vec{u})dt = 0. \qquad (2.61)$$

Since no work is done, a magnetic field cannot change the kinetic energy of a charged particle; *the magnetic field can change the direction of motion of a charged particle but it cannot change its speed.*

Up to this point, the magnetic flux density $\vec{B}$ was used to indicate the presence of a magnetic field in a given region of space. Now, we are going to use the magnetic field intensity, $\vec{H}$. For most materials $\vec{B}$ and $\vec{H}$ are linearly related (see Eq. (2.57)), therefore knowledge of one is sufficient.

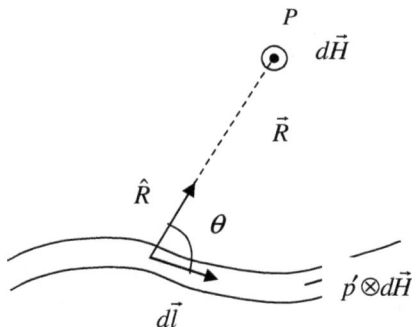

Fig. 2.13. Magnetic field $d\vec{H}$ generated by a current element $I d\vec{l}$ ($d\vec{H}$ points into the page).

Through his experiments on the deflection of the compass needle by current-carrying wires, Hans Oersted established that currents induce magnetic fields that form closed loops around the wires. On the basis of Oersted results, Jean Biot and Felix Savart derived an expression that relates the magnetic field $\vec{H}$ at any point in space to the current I that generates $\vec{H}$. The Biot–Savart Law states that the differential magnetic field $d\vec{H}$ generated by a steady current I flowing through a differential length $d\vec{l}$ is given by:

$$dH = \frac{I}{4\pi} \frac{d\vec{l} \times \hat{R}}{R^2} \, \text{A/m}, \qquad (2.62)$$

where $\vec{R} = \hat{R}R$ is the distance vector between $d\vec{l}$ and the observation point P shown in Fig. 2.13 and $\hat{R}$ is the unit vector.

The direction of the field induced at point P is opposite to the field induced at point P'. The SI unit for $\vec{H}$ is ampere $\cdot$ m/m$^2 = $ A/m. It is important to remember that the direction of the magnetic field is defined such that $d\vec{l}$ is along the direction of the current I and the unit vector $\hat{R}$ points *from* the current element to the observation point.

According to Eq. (2.62), $d\vec{H}$ varies as R^{-2}, which is similar to the distance dependence of the electric field induced by an electric charge. However, unlike the electric field vector $\vec{E}$, whose direction is *along* the distance vector $\vec{R}$ connecting the charge to the observation point, the magnetic field $\vec{H}$ is *orthogonal* to the plane containing the direction of the current element $d\vec{l}$ and the distance vector $\vec{R}$. At point P in Fig. 2.13, the direction of $d\vec{H}$ is out of page, whereas at point P' the direction of $d\vec{H}$ is into the page.

In order to determine the total magnetic field $\vec{H}$ due to a conductor of a finite size, we need to add the contributions of all the current elements comprising the conductor. Hence, the Biot–Savart Law becomes:

$$\vec{H} = \frac{I}{4\pi} \int_l \frac{d\vec{l} \times \hat{R}}{R^2} \, \text{A/m}, \qquad (2.63)$$

where l is the line along which I exists.

We will now examine Eq. (2.56), which is the second of Maxwell's pair of equations characterizing the magnetostatic fields, $\vec{B}$ and $\vec{H}$.

$$\nabla \times \vec{H} = \vec{J}. \qquad (2.64)$$

The integral form of (2.64) is called *Ampere's Law* or Ampere's circuital law under magnetostatic conditions (steady currents). It is obtained by integrating both sides of (2.64) over the open surface S.

$$\int_S (\nabla \times \vec{H}) \cdot d\vec{S} = \int_S \vec{J} \cdot ds \qquad (2.65)$$

and then using Stoke's theorem Eq. (2.66):

$$\int_S (\nabla \times \vec{B}) \cdot d\vec{s} = \oint_C \vec{B} \cdot d\vec{l} \quad \text{(Stoke's theorem)} \qquad (2.66)$$

to obtain the result:

$$\oint_C \vec{H} \cdot d\vec{l} = I \quad \text{(Ampere's Law)}, \qquad (2.67)$$

where C is the closed contour bounding the surface S. *The sign convention for the direction of C is taken so that $\vec{I}$ and $\vec{H}$ satisfy the right-hand rule* defined earlier in connection with the Biot–Savart Law. That is, if the direction of I is aligned with the direction of the thumb of the right hand, then the direction of the contour C should be chosen to be along the direction of the other four fingers.

Ampere's (circuital) Law states that the line integral of $\vec{H}$ around a closed path is equal to the current traversing the surface bounded by that path. In Fig. 2.14, the line integral of $\vec{H}$ is equal to the current I, even though the paths have different shapes and the magnitude of $\vec{H}$ is not uniform along the path of configuration (b). By the same token, since path (c) in Fig. 2.14 does not enclose the current I, its line integral of $\vec{H}$ is identically zero, even though $\vec{H}$ is not zero along the path.

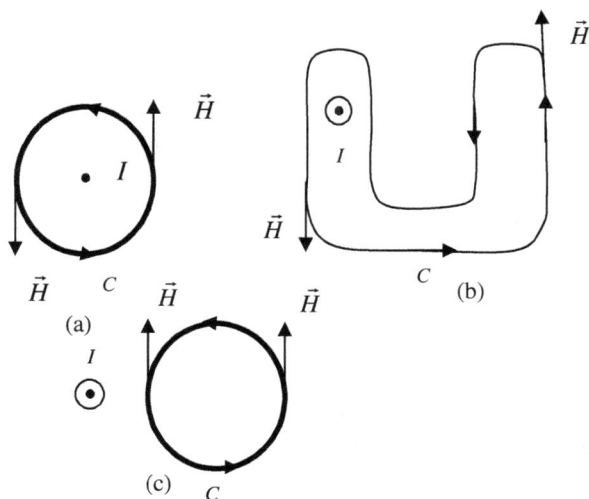

Fig. 2.14. Ampere's Law: (a, b) Line integral of $\vec{H}$ around a closed contour C is equal to the current traversing the surface bounded by the contour; (c) Current I is not enclosed by the contour C, therefore the line integral of $\vec{H}$ is zero.

When we considered Gauss's Law earlier in the chapter, we determined that its usefulness for calculating the electric flux density $\vec{D}$ is limited to charge distributions that possess a certain degree of symmetry and the calculation procedure is subject to proper choice of the Gaussian surface enclosing the charges. A similar statement applies to Ampere's Law: its usefulness is limited to symmetric current distributions that allow the choice of convenient Ampere's contours around them.

2.14. Dynamic fields

Electric charges induce electric fields and electric currents induce magnetic fields. As long as the charge and current distributions remain constant in time, so will the fields they induce. If the current varies with time t, not only will the fields vary with time as well but the electric and magnetic fields become interconnected, and the coupling between them produces electromagnetic waves capable of traveling through free space and in material media. In order to study time-varying electromagnetic phenomena, we need to use Maxwell's equations as an integrated unit. These equations are given in both differential and integral form in:

Whereas in the static case ($\delta/\delta t = 0$), we were able to use the first pair of Maxwell's equations for the electrostatics and magnetostatics. In the

Table 2.7. Maxwell's equations.

Gauss's Law	$\nabla \cdot \vec{D} = \rho_v$	$\oint_S \vec{D} \cdot d\vec{s} = Q$	(2.68)
Faraday's Law	$\nabla \times \vec{E} = -\dfrac{\delta \vec{B}}{\delta t}$	$\oint_C \vec{E} \cdot d\vec{l} = -\displaystyle\int_S \dfrac{\delta \vec{B}}{\delta t} \cdot d\vec{s}$	(2.69) (stationary surface S)
Gauss's Law for magnetism (no magnetic charges)	$\nabla \cdot \vec{B} = 0$	$\oint_S \vec{B} \cdot d\vec{s} = 0$	(2.70)
Ampere's Law	$\nabla \times \vec{H} = \vec{J} + \dfrac{\delta \vec{D}}{\delta t}$	$\oint_C \vec{H} \cdot d\vec{l} = \displaystyle\int_S \left(\vec{J} + \dfrac{\delta \vec{D}}{\delta t} \right) \cdot d\vec{s}$	(2.71)

dynamic case, we have to deal with the coupling that exists between the electric and magnetic fields as expressed by the second and fourth equations in Table 2.7. The first equation represents Gauss's Law, and it is equally valid for static and dynamic fields. The same is true for the third equation $\nabla \cdot B = 0$ that means there are no magnetic charges. The second and fourth equations have different meanings for static and dynamic fields. In the dynamic case, a time-varying magnetic field gives rise to an electric field (Faraday's Law) and, vice-versa, a time-varying electric field produces the results and expressions for the fields under dynamic conditions will reduce to those applicable under static conditions.

The connection between electricity and magnetism was established by Oersted, who demonstrated that a wire carrying an electric current exerts a force on a compass needle and that the needle always turn so as to point in the $\hat{\phi}$-direction when the current is along the $\hat{z}$-direction. The force acting on the compass needle is due to the magnetic field produced by the current in the wire. However, to induce an electric current in a wire, the magnetic field should change with time. In order to explain how the induction takes place, let us consider the set-up in Fig. 2.15.

A square loop made of a conducting wire is connected to a galvanometer that is placed next to a conducting coil. The coil is hooked up to a battery. The current in the coil produces a magnetic field $\vec{B}$ whose lines pass through the loop as shown in Fig. 2.15. The magnetic flux Φ passing through a loop is the integral of the normal component of the magnetic flux density over the surface area of the loop, S:

$$\Phi = \int_S \vec{B} \cdot d\vec{s} \ \text{Wb}. \qquad (2.72)$$

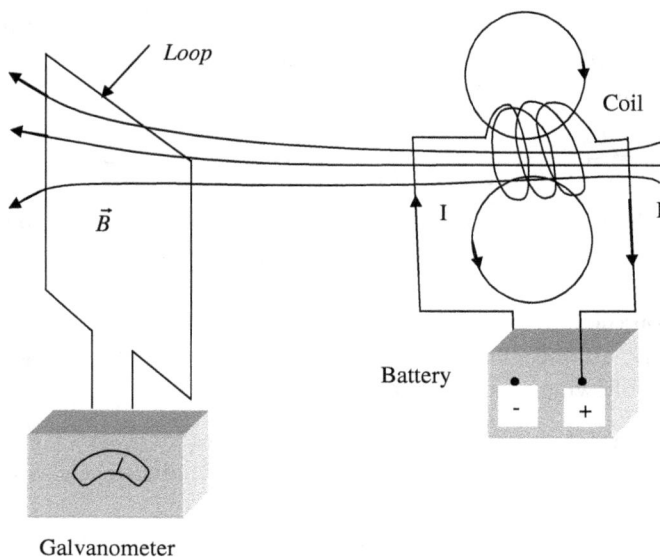

Fig. 2.15. Galvanometer showing a deflection whenever the varying with time magnetic flux passes through the square loop.

Under stationary conditions, the $d - c$ current in the coil produces a constant magnetic field $\vec{B}$, which in turn produces a constant flux through the loop. When the flux is constant, no current is detected by the galvanometer. However, when the battery is disconnected, thereby interrupting the current flow in the coil, the magnetic field drops to zero, and the consequent change in the magnetic flux will cause a momentary deflection of the galvanometer needle. When the battery is hooked up again, the galvanometer will exhibit a momentary deflection but in the opposite direction. Thus, current is induced in the loop when the magnetic flux changes, and the direction of the current depends on whether the flux is increasing (battery is connected) or decreasing (the battery is disconnected). It was further determined that current can also flow in the loop, while the battery is connected to the loop, and the loop is turned around with acceleration or if we move the loop farther or closer to the coil. The physical movement of the loop changes the amount of flux linking its surface S, even though the field $\vec{B}$ due to the coil has not changed.

The galvanometer is the predecessor of the voltmeter and ammeter. When a galvanometer detects a flow of current through the coil, it means that a voltage has been induced across the galvanometer terminals. This voltage is called the electromagnetic force (emf), V_{emf}, and the process is called *electromagnetic induction*. The *emf* induced in a closed conducting loop of N turns is given by:

$$V_{\text{emf}} = -N\frac{d\Phi}{dt} = -N\frac{d}{dt}\int_S \vec{B} \cdot d\vec{s} \; V \quad \text{(Faraday's Law).} \qquad (2.73)$$

Even though the results leading to Eq. (2.73) were discovered by both Henry and Faraday, (2.73) is known as *Faraday's Law*.

Note that the derivative in (2.73) is a total time derivative that operated on the magnetic field $\vec{B}$, as well as the differential surface area $d\vec{s}$. Accordingly, an emf can be generated in a closed conducting loop under any of the following three conditions:

1. A time-varying magnetic field links two loops (*transformer emf*): V_{emf}^{tr}.
2. A moving loop with a time varying area, the induced emf is called the *motional emf*, V_{emf}^{m}.
3. A *moving loop* in a *time-varying field B*. The total emf is given by:

$$V_{\text{emf}} = V_{\text{emf}}^{tr} + V_{\text{emf}}^{m} \qquad (2.74)$$

— for the stationary (Case 1), $V_{\text{emf}}^{m} = 0$;
— if $\vec{B}$ is static (Case 2), then $V_{\text{emf}}^{tr} = 0$;
— neither term is zero;

As we have already established, a changing magnetic flux induces an electric field that is described by Eq. (2.75) in the form:

$$\oint \vec{E} \cdot d\vec{s} = -\frac{d\Phi_B}{dt} \quad \text{(\textit{Faraday's Law of induction}).} \qquad (2.75)$$

Here $\vec{E}$ is the electric field induced along a closed loop by the changing magnetic flux Φ_B encircled by that loop. Since symmetry is often so powerful in physics, we are tempted to guess whether induction can occur in the opposite sense; that is, can a changing electric flux induce a magnetic field?

The answer is yes, it can; furthermore, the equation governing the induction of a magnetic field is almost symmetric with Eq. (2.75). It is often called Maxwell's Law of induction after James Clerk Maxwell:

$$\oint \vec{B} \cdot d\vec{s} = \mu_0\varepsilon_0\frac{d\Phi_E}{dt} \quad \text{(\textit{Maxwell's Law of induction}).} \qquad (2.76)$$

Here $\vec{B}$ is the magnetic field induced along a closed loop by the changing electric flux Φ_E in the region encircled by that loop.

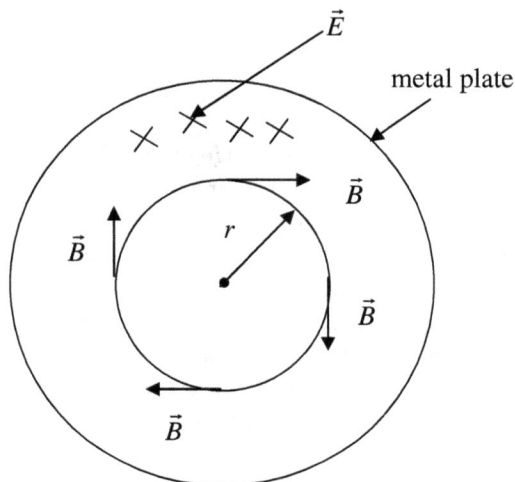

Fig. 2.16.　Electric field is uniform and directed into the page and grows in magnitude as the charge on the plate increases.

Although Eq. (2.76) is similar to Eq. (2.75), the equations differ in two ways. First, (2.75) has the two extra symbols μ_0 and ε_0, but they appear only because we employ SI units. Second, (2.76) lacks the minus sign of (2.75), meaning that the induced electric field $\vec{E}$ and the induced magnetic field $\vec{B}$ have opposite directions when they are produced in otherwise similar situations. To see why they are different, examine the situation when an increasing magnetic field $\vec{B}$ induces an electric field $\vec{E}$. The induced field $\vec{E}$ is counterclockwise, opposite the induced magnetic field $\vec{B}$ (Fig. 2.16).

Now, recall that the left side of Eq. (2.76), the integral of the dot product $\vec{B} \cdot d\vec{s}$ around a closet loop, appears in another equation — namely, Ampere's Law:

The magnetic field $\vec{B}$ induced by this changing electric field is shown at four points on a circle with a radius r, less than the plate's radius

$$\oint \vec{B} \cdot d\vec{s} = \mu_0 i_{\text{ecl}} \quad \text{(Ampere's Law)}, \tag{2.77}$$

where i_{enc} is the current encircled by the closed loop. Thus, our two equations that specify the magnetic field $\vec{B}$ produced by means other than a magnetic material (that is, by a current and a changing electric field) give the field in exactly the same form. We can combine the two equations into the single equation:

$$\oint \vec{B} \cdot d\vec{s} = \mu_0 \varepsilon_0 \frac{d\Phi_E}{dt} + \mu_0 i_{\text{enc}} \quad \text{(Ampere–Maxwell's Law)}. \qquad (2.78)$$

When there is a current but no change in the electric flux (such as with a wire carrying a constant current), the first term on the right side of (2.78) is zero, and so Eq. (2.78) reduces to Eq. (2.77), Ampere's Law. When there is a change in the electric flux but no current (such as inside or outside the gap of a charging capacitor), the second term on the right side of Eq. (2.78) is zero, and so Eq. (2.78) reduces to Eq. (2.76), Maxwell's Law of induction.

If you compare the two terms on the right side Eq. (2.78), you will see that the product $\varepsilon_0 (d\Phi_E/dt)$ must have the dimension of a current. In fact, that product has been treated as being a fictitious current called the *displacement current* (Ravand and Lemarquand, 2009):

$$i_d = \varepsilon_0 \frac{d\Phi_E}{dt} \quad (displacement \ current). \qquad (2.79)$$

The word *"displacement"* is poorly chosen in that nothing is being displaced, but we are stuck with the term. Nevertheless, we can now rewrite Eq. (2.78):

$$\oint \vec{B} \cdot d\vec{s} = \mu_0 i_{\text{d,enc}} + \mu_0 i_{\text{enc}} \quad \text{(Ampere–Maxwell's Law)}, \qquad (2.80)$$

where $i_{\text{d,enc}}$ is the displacement current that is encircled by the integration loop.

Let us consider a charging capacitor with circular plates, as in Fig. 2.17.

The real current i that is charging the plates changes the electric field $\vec{E}$ between the plates. The fictitious displacement current i_d between the plates is associated with that changing field $\vec{E}$. Let us relate these two currents.

The charge q on the plates at any time is related to the magnitude E of the field between the plates at that time by Eq. (2.81):

$$q = \varepsilon_0 A E, \qquad (2.81)$$

where A is the plate area. To determine the real current i, we differentiate Eq. (2.81) with respect to time, finding:

$$\frac{dq}{dt} = i = \varepsilon_0 A \frac{dE}{dt}. \qquad (2.82)$$

To receive the displacement current i_d, we can use Eq. (2.79). Assuming that the electric field $\vec{E}$ between two plates is uniform (we neglect any

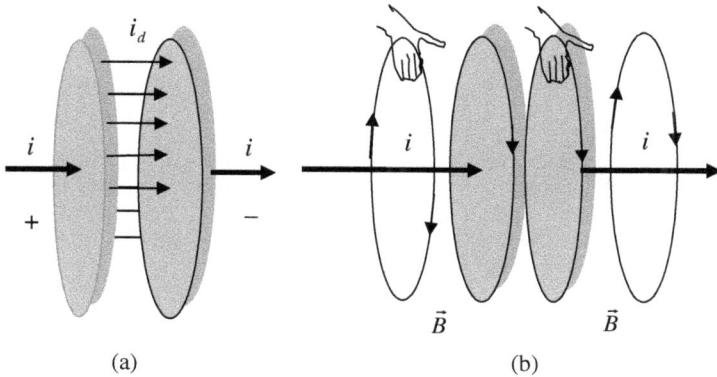

(a)   (b)

Fig. 2.17.   (a) The displacement current i_d between the plates of a capacitor that is being charged by a current I; (b) The right-hand rule for finding the direction of the magnetic field around a wire with a real current (as on the left) also gives the direction of the magnetic field around a displacement current (as in the center).

fringing), we can replace the electric flux Φ_E in that equation with EA. Then, Eq. (2.79) becomes:

$$i_d = \varepsilon_0 \frac{d\Phi_E}{dt} = \varepsilon_0 \frac{d(EA)}{dt} = \varepsilon_0 A \frac{dE}{dt}. \qquad (2.83)$$

Comparing (2.82) and (2.83), we see that the real current i charging the capacitor and the fictitious displacement current i_d between the plates have the same magnitude:

$$i_d = i \quad \text{(displacement current in a capacitor)}. \qquad (2.84)$$

Thus, we can consider the fictitious displacement current i_d to be simply a continuation of the real current I from one plate, across the capacitor gap, to the other plate. Since the electric field is uniformly spread over the plates, the same is true of this fictitious displacement current i_d, as suggested by the spread of current arrows in Fig. 2.17(a). Although no charge actually moves across the gap between the plates, the idea of the fictitious current i_d can help us to quickly find the direction and magnitude of an induced magnetic field.

References

Ravand, R and G Lemarquand (2009). Introducing fictitious currents for calculating analytically the electric field in cylindrical capacitors. *Progress in Electromagnetic Research*, 9, 139–150.

Chapter 3

Mathematical Methods of Identification

Computational methods have been in use almost as long as the history of humanity can be traced. Today, complicated technologies require complicated computational algorithms and operations. The difficulty in THz measurements dictates sophisticated methods of evaluation and even prediction of data. With still insufficient power of THz sources, missing data and scarce representation are major problems. Below is a short description of three possible choices for THz identification computational methods. The mathematical technique suggested by the author has features of several numerical and analytical approaches. It is basically time series analysis (TSA) modified for a number of particular purposes. The validity of this approach has been tested in several projects carried out by the author.

Both, the wavelet analysis (see for more details Ch. 8) and perturbation theory (PT) are applicable to THz identification problems, however, none of them can predict a possible outcome that delays or even indefinitely postpones the recognition of the object until the analysis is performed and the results are classified and decided upon. It is more practical, however, to try to foresee the appearance of a possible dangerous item and create an approximate image of it. Then, we can compare the received image of the target with a mathematical model based on the previous measurements with the real image's likely deviations from the model taken into account — that is what TSA does. As mentioned before, parts of the wavelet and perturbation theories were included in the proposed approach. In particular, autoregressive time series models are close to the PT gradual approximation, the elements of which may be substituted and corrected. Time series analysis also shares the adaptivity and *ad hoc* nature of wavelet analysis with the additional ability to model and predict the characteristics of a potential object. Below are the main features of the time series, and wavelet analyses and PT.

The natural ordering found in TSA distinguishes it from other common data analysis problems which have no natural order of observation. TSA is also distinct from spatial data analysis where the observations typically relate to their spatial location. Time series generally reflect the fact that the observations close together in time will be more closely related than observations farther apart. In addition, time series models often make use of the natural one-way ordering of time so that values for a given period are expressed as being derived from past values, rather than from future ones. However, it is the prediction that is the main objective of TSA.

As a mathematical tool, wavelet analysis can be used to extract information from many different kinds of data, particularly images. Sets of wavelets are generally needed to analyze the data fully. Sets of complementary wavelets are also used in compression and decompression algorithms for extracting information from the signal.

In general, a wavelet is similar to a wave-like oscillation whose amplitude increases from zero to a maximum and then goes back to zero again. The wavelets have resonant sensitivity to the signal's frequency, i.e. the wavelet can determine a specific frequency pitch that makes the wavelet useful for identification. Wavelet analysis often implies complicated combinations of operations that help to extract segments of information. All wavelet transforms may be considered as forms of time–frequency representation for analog signals and are thus related to harmonic analysis. Almost all practically useful discrete wavelet transforms use discrete-time filter banks. These filter banks are called the wavelet coefficients. These filter banks may contain either finite impulse response (FIR) or infinite impulse response (IIR) filters. The wavelets forming a continuous wavelet transform (CWT) are subject to the uncertainty principle of Fourier analysis respective sampling theory: Given a signal, it is impossible to simultaneously, assign an exact time and frequency response scale to an event in the signal. The product of the uncertainties of time and frequency response scale has a lower limit. Therefore, in the scalogram of a continuous wavelet transform of this signal, such an event marks an entire region in the time-scale plane, instead of just one point.

PT comprises of mathematical methods that are used to find an approximate solution to a problem which cannot be solved exactly, by starting from the exact solution of a related problem. PT is applicable if the solution may be found by adding a small term to the mathematical model of the exact solution. PT leads to an expression for the desired solution in terms of a power series in small parameters, which forms PT itself. The leading term in this power series is the solution of the problem that may be solved exactly, while further terms describe the deviation in the solution. Thus, we have an approximation to the full solution (A) in a series of small parameters:

$$A = A_0 + \varepsilon^1 A_1 + \varepsilon^2 A_2 + \cdots ,$$

where A_0 is the exact solution and $A_1 \ldots A_n$ represent the higher-order terms that may be found iteratively by a systematic procedure. An approximate solution for PT is obtained by eliminating the smaller terms. For example, the first order perturbation series:

$$A \approx A_0 + \varepsilon A_1.$$

Perturbation methods start with a simplified form of the original problem, which is simple enough to be solved exactly. The slight changes that result from accommodating the perturbation (that may be simplified again) are used to correct the approximate solution. Because of the simplification at each step, the resulting solution is never perfect. However, often even several steps of approximations provide an acceptable solution. There is no requirement to stop at any of the steps. Thus, a partially corrected solution can be re-used as the new starting point for another cycle of approximations. The difficulty is often the growing complexity of the problem.

TSA is one way to build a mathematical model for the above goal, and to predict, reconstruct, correct or complement the image of an object (Chatfield, 1989). These are two main goals of TSA: (a) To identify the nature of the phenomenon represented by the sequence of observations, and (b) To predict future values of observed time series data that are identified and formally described (forecast). Once a preliminary mathematical model is created, we can interpret and fill it with data (and thus use it as our theoretical model of the investigated phenomenon). Regardless of the depth of our understanding of the problem and the validity of our theoretical approach of the phenomenon, we can extrapolate the created model to predict future events.

As in many mathematical analyses, in TSA it is assumed that the data consist of a systematic pattern (usually a set of identifiable components) and random noise (error) that usually makes the pattern difficult to identify. Most TSA techniques involve some form of noise filtering in order to clarify the pattern. (Gonzalez and Woods, 1992; Jain, 1989; Marion, 1999).

The term TSA is used to distinguish a problem, first of all from more ordinary data analysis problems (where there is no natural order of the context of individual observations), and secondly from spatial data analysis where observations often relate to subjects in space (e.g. geographical position). These are additional possibilities in the form of space–time models (often called spatial-temporal analysis). A time series model will generally reflect the principle that similar events or similar characteristics of events will yield a prediction of events or characteristics that are also similar. In addition, time series models will often make use of the existing order in which events take place in time so that values in a series for a given moment will be expressed as being derived from past values rather than from future values.

Methods for TSA are usually divided into two classes: frequency-domain methods and time-domain methods. The former centers on spectral analysis and wavelet analysis (Donald, 2000), and can be regarded as model-free analysis well-suited to inquiry-like investigations. Time-domain methods have a model-free subset consisting of autocorrelation and cross-correlation analyses — they are the basis for specified time series models.

Models for time series data can have many forms and represent different stochastic processes. It is worthwhile to mention here the wide usage of the notion "stochastic process" for computational prediction techniques. A stochastic process (sometimes called "random process") is a collection of variables that is used to represent the evolution of some random value, or system in time. This is the probabilistic counterpart to a deterministic process. Instead of describing a process that can only evolve one way, in a stochastic process, there is some indeterminacy: even with the same initial condition, there are different ways in which the process may evolve. The term "random", does not necessarily mean the absolutely chaotic process, it usually only means that we are dealing with some probability of an event.

There are three broad classes of practical importance that deal with model variations: the autoregressive (AR) models, the integrated (I) models, and the moving average (MA) models. These three classes depend linearly on previous data points. Combinations of these classes produce *autoregressive moving average* ($ARMA$) and *autoregressive integrated moving average* ($ARIMA$) models. The *autoregressive fractionally integrated moving average* ($ARFIMA$) model generates the former three. We can extend the above classes to deal with vector-valued data. Multivariate time series models are extended by including an initial "V" for "vector". An additional set of extensions of these models is available for use where the observed time series is generated by some *"forcing"* time series (which may not have a casual effect on the observed series): the distinction from the multivariate case is that the forcing series may be deterministic or under the experimenter's control. For these models, the acronyms are extended with a final *"X"* for *"exogenous"*.

Nonlinear dependence of the level of a series on previous data points has some significance, partially because of the possibility of producing a *chaotic* time series. Chaotic time series are related to an area in mathematics and physics that studies the behavior of certain dynamical systems, which are highly sensitive to initial conditions. The deterministic nature of the subjects of THz identification does not mean that all their physical parameters are predictable (Kellert, 1993). As it was mentioned at the beginning of the chapter, many physical parameters of the subject change with time or depending on circumstances. Thus, a possible inclusion of chaos theory in the model building process is a promising way to increase the reliability of the identification. Also, from practical applications we can see the advantage of using predictions derived from nonlinear models, over those from linear models.

Among other types of nonlinear time series models, these are models to represent the changes of variance with time (heteroskedasticity) (Mills, 1990; Percival and Walden, 1993; Pandit and Wu, 1983). In statistics, a sequence of random variables is *heteroskedastic* or *heteroscedastic* if the random variables have different variances. In contrast, a sequence of random variables is called *homoscedastic* if it has constant variance. Now we need to make a number of assumptions: One of them is that the error term has a constant variance. It is important to take into account that in general the variance is not constant. That is why, heteroskedisity becomes essential while developing mathematical models with increased accuracy and reliability. These models are called autoregressive conditional heteroskedasticity ($ARCH$) (Hamilton, 1994) and there is a wide variety of representation ($GARCH,\ TARCH,\ EGARCH,\ FIGARCH,\ CGARCH,$ etc.).Here, changes in variability are related to, or predicted by, recent past values of the observed series. On the other hand, there may be other possible representations of locally-varying variability, where the variability might be modeled as being generated by a separate time-varying process, as in a doubly stochastic model.

One of the possible models that can be used for identification model building is an *autoregressive (AR) model.* In statistics and signal processing, an autoregressive model is one of a group of linear prediction formulas that attempt to predict an output of a system based on previous outputs. Here, we can use a library of the known physical parameters of a subject to refer to as a source of previous outputs.

AR is a model that has no trend (the constant mean is taken as 0). Let $X_1, X_2 \ldots$ be successive instances of the random variable X measured at regular intervals of time. Let ε_j be the random variable denoting the random error at time j. As pth-order autoregressive model (or autoregressive process) relates the value at time j to the preceding p values by $X_j = a_1 X_{j-1} + a_2 X_{j-2} + \cdots + a_p X_{j-p} + \varepsilon_j$, where $a_1, a_2, \ldots, a_p$ are constants. Such a model is written in brief as $AR(p)$.

Thus, the notation $AR(p)$ refers to the autoregressive model of order p. Alternatively, the $AR\,(p)$ model is defined as:

$$X_t = c + \sum_{i=1}^{p} \varphi_i X_{t-i} + \varepsilon_t, \tag{3.1}$$

where $\varphi_1, \ldots, \varphi_p$ are the *parameters* of the model, c is a constant and ε_t is *white noise* (i.e. a random signal or process with a flat power spectral density. In other words, the signal has equal power within a fixed bandwidth of this signal). The constant component is omitted by many authors for simplicity. Also, for the sake of simplicity, we can think about an all-pole infinite impulse response filter with some additional interpretation associated with it. *All-pole filter* is a filter that has a frequency response that goes to infinity at specific frequencies. All-pole modeling is used in, for e.g. signal processing (in particular in linear prediction) (Backstrom and Alku, 2003). Some constraints should be placed on the values of the parameters of this model in order that the model remains stationary. A *stationary process* is a stochastic process whose joint probability distribution does not change when shifted in time or space. As a result, parameters such as *mean* and *variance*, if they exist, do not change over time or position. Stationarity is used as a tool in TSA, where the raw data are often transformed to become stationary. For example, processes in the $AR(1)$ model with $|\varphi_1| \geq 1$ are not stationary.

Example: *An AR(1) — process*

An AR(1) — process is given by:

$$X_t = c + \varphi X_{t-1} + \varepsilon_t, \tag{3.2}$$

where c is a constant, ε_t is a white noise process with zero mean and variance σ^2. (*Note*: the subscript on φ_1 has been dropped). White noise is a random signal (or process) with a flat power spectral density (i.e. the signal contains equal power within a fixed bandwidth at any center frequency). The process qualifies as wide-sense stationarity (WSS) if $|\varphi < 1|$. (If $\varphi = 1$ then X_t exhibits a unit root and can also be considered as a random walk, which is not WSS.) Assuming $|\varphi < 1|$ and denoting the mean by μ, we receive:

$$E(X_t) = E(c) + \varphi E(X_{t-1}) + E(\varepsilon_t)$$
$$\Rightarrow \mu = c + \varphi\mu + 0.$$

Thus,

$$\mu = \frac{c}{1 - \varphi}. \tag{3.3}$$

In particular, if $c = 0$, then the mean is 0. The variance can be shown to be equal:

$$\text{var}(X_t) = E(X_t^2) - \mu^2 = \frac{\sigma^2}{1 - \varphi^2} \varphi^{|n|}. \tag{3.4}$$

The autocovariance (i.e. how much X changes in time with respect to itself) is given by:

$$B_n = E(X_{t+n} X_t) - \mu^2 = \frac{\sigma^2}{1 - \varphi^2} \varphi^{|n|}. \tag{3.5}$$

It can be seen that the autocavariance function decays with a decay time (also called *time constant*) of $\tau = -1/\ln(\varphi)$ [to see this, write $B_n = K\varphi^{|n|}$, where K is independent of n].

Then, note that $\varphi^{|n|} = e^{n \ln \varphi}$ and match this to the exponential decay law $e^{-n/\tau}$.

The spectral density function is the Fourier transform of the auto-covariance function. In discrete terms, this will be discrete-time Fourier transform:

$$\Phi(\omega) = \frac{1}{\sqrt{2\pi}} \sum_{n=-\infty}^{\infty} B_n e^{-i\omega n} = \frac{1}{\sqrt{2\pi}} \left(\frac{\sigma^2}{1 + \varphi^2 - 2\varphi \cos(\omega)} \right). \tag{3.6}$$

This expression is periodic due to the discrete nature of the X_j, which is manifested as the cosine term in the denominator. If we assume that the sampling time $\Delta t = 1$ is much smaller than the decay time (τ), then we can use a continuum approximation to B_n:

$$B(t) \approx \frac{\sigma^2}{1 - \varphi^2} \varphi^{|t|},$$

(3.7)

which yields a Lorentzian profile for the spectral density:

$$\Phi(\omega) \approx \frac{1}{\sqrt{2\pi}} \frac{\sigma^2}{1 - \varphi^2} \frac{\gamma}{\pi(\gamma^2 + \omega^2)},$$

(3.8)

where $\gamma = 1/\tau$ is the angular frequency associated with the decay time τ. An alternative expression for X_t can be derived by first substituting $c + \varphi X_{t-2} + \varepsilon_{t-1}$ for X_{t-1} in the defining equation. Continuing this process N times yields:

$$X_t = c \sum_{k=0}^{N-1} \varphi^k + \varphi^N X_{t-N} + \sum_{k=0}^{N-1} \varphi^k \varepsilon_{t-k}.$$

(3.9)

For N approaching infinity, φ^N will approach zero and

$$X_t = \frac{c}{1 - \varphi} + \sum_{k=0}^{\infty} \varphi^k \varepsilon_{t-k}.$$

(3.10)

It is seen that X_t is white noise convolved with the φ^k *kernel* plus the constant mean. A kernel is a weighting function used in nonparametric estimation techniques. Kernels are used in kernel density estimation to evaluate random variables' density functions, or in kernel regression to estimate the conditional expectation of a random variable. Here, we use the kernel to estimate spectral density.

If the white noise ε_t is a Gaussian process (i.e. a stochastic process $\{X_t\}_{t \in T}$ for which any finite linear combination of samples will be normally distributed) then X_t is also a Gaussian. In other cases, the central limit theorem indicates that X_t will be approximately normally distributed when φ is close to one.

Calculation of the AR parameters:

The $AR(p)$ model is given by the equation:

$$X_t = \sum_{i=1}^{p} \varphi_i X_{t-i} + \varepsilon_t.$$

(3.11)

It is based on parameters φ_i where $i = 1, \ldots, p$. There is a direct correspondence between these parameters and the covariance function of the process, and this correspondence can be inverted to determine the parameters from the autocorrelation function (which is itself obtained from the covariances). This is done using *Yule–Walker* equations:

$$\gamma_m = \sum_{k=1}^{p} \varphi_k \gamma_{m-k} + \sigma_\varepsilon^2 \delta_m, \qquad (3.12)$$

where $m = 0, \ldots, p$, yielding $p + 1$ equations. γ_m is the autocorrelation function of X, σ_ε is the standard deviation of the input noise process, and δ_m is the Kronecker delta function (i.e. a function of two variables. It equals unity if the variables are equal, and zero if otherwise).

Since the last part of the equation is nonzero only if $m = 0$, the equation is usually solved by representing it as a matrix for $m > 0$, thus receiving:

$$\begin{bmatrix} \gamma_1 \\ \gamma_2 \\ \gamma_3 \\ \vdots \end{bmatrix} = \begin{bmatrix} \gamma_0 & \gamma_{-1} & \gamma_{-2} & \cdots \\ \gamma_1 & \gamma_0 & \gamma_{-1} & \cdots \\ \vdots & \vdots & \vdots & \ddots \end{bmatrix} \begin{bmatrix} \varphi_1 \\ \varphi_2 \\ \varphi_3 \\ \vdots \end{bmatrix} \qquad (3.13)$$

solving all φ. Here, M is a general form of a variable in the matrix, i.e. $\gamma_p, \gamma_{p-1} \ldots$ For $m = 0$ we have:

$$\gamma_0 = \sum_{k=1}^{p} \varphi_k \gamma_{-k} + \sigma_\varepsilon^2, \qquad (3.14)$$

which allows us to solve σ_ε^2. The above *Yule–Walker* equations provide one route to estimate the parameters of an $AR(p)$ model, by replacing the theoretical covariances with estimated values. One way to specify the estimated covariences is a calculation using *least squares regression* of values X_t on the previous p values of the same series. Linear regression models are often fitted using the least squares approach. Conversely, the least squares approach can also be used to fit nonlinear models. Thus, while the terms *least squares* and *linear models* are closely related, they are not synonymous.

In statistics, *linear regression* refers to any approach to modeling the relationship between one or two variables denoted y and one or more variables denoted X, such that the model depends linearly on the unknown parameters to be estimated from the data. Linear regression has many practical applications. Most of the applications of linear regression suitable for our discussion fall into one of the following two categories:

(1) If the goal is prediction, or forecasting, linear regression can be used to fit a predictive model to an observed data set of y and X values. After developing such a model, if an additional value of X is given without its accompanying value of y, then the fitted model can be used to make a prediction of the value y.

(2) Given a variable y and a number of variables $X_1, \ldots, X_p$ that may be related to y, linear regression analysis can be applied to quantify the strength of the relationship between y and X_j to assess which subsets of X_j contain redundant information about y. Thus, once one of them is known, the others are no longer informative. In Sec. 3.1, we will show how methods of linear regression may be applied to concrete problems of THz identification.

The principle of identification relies on permanent features that are certain in the memory of the identification devices. In time series, there are two main components that are called *trend* and *seasonality*. It must be pointed out that though both components may have similar natures, the latter repeats itself over time, while the former, at the same time, represents a general systematic linear or (most often) nonlinear component that changes over time and does not repeat or at least does not repeat within the time range captured by the data. The latter may have a formally similar nature; however, it repeats itself in systematic intervals over time.

The next aspect of identification is the case when we do not know what the object is exactly, nor is its description written in the memory of the identification apparatus. When the patterns of the data are unclear, individual observations involve considerable error, and we still need not only to uncover the hidden patterns in the data but also generate forecasts. For developing forecasts, a number of mathematical methods and procedures exist (e.g. autoregressive process, stationary requirement, moving average process, autoregressive moving average models, etc.)

In summary, the discourse presented above argues that there are two distinct cases of identification: (1) when the received images coincide (with some error) with one of the library (predetermined) images or (2) when they do not coincide. The second case is more complicated and provides an image or a description of a possible object with a degree of correlation to the known objects or their parameters. For example, if the identified substance does not fall exactly into one of the library descriptions but there is a degree of correlation with them, then the computer system provides a possible type of the substance in question.

Further in the chapter, the mathematical methods usable for identification are elaborated on along with applicable examples.

3.1. Process modeling

3.1.1. *General approach*

Once the input signal is received, detected and amplified, it is analyzed which, in essence, means building a mathematical model of it. The signal is divided into several major *deterministic* and *random* components. The deterministic component is given by a mathematical function: x_1, x_2, etc. The random component follows a particular probability distribution. The deterministic component is usually determined from a library of well-established patterns and tendencies (e.g. such as one produced by an electromagnetic field interacting with the identified objects) or assumed based on previous measurements or events.

In order to choose the modeling process, let us consider a process with only one explanatory variable. Let us introduce a common situation when an identification system's probing beam encounters a moving object. Figure 3.1 depicts a distribution of the measured intensity of an electromagnetic beam at the receiver as a function of time. The intensity deviation resembles a descending sine that corresponds to the oscillatory

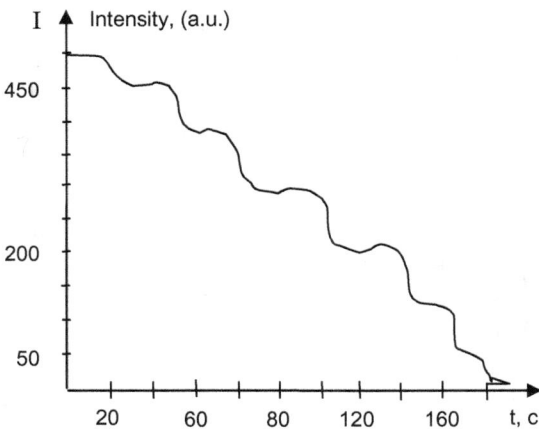

Fig. 3.1.   Variation of intensity versus time.

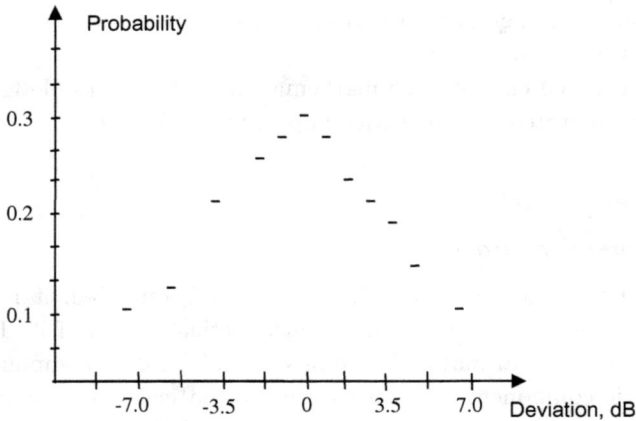

Fig. 3.2.   Point-diagram of probability of deviation from the general curve.

character of various processes. The intensity drops with time (e.g. probing beam attenuated by fog) but the current maintains its periodicity. I is the intensity at the receiver (Sokolnikov, 2006a).

In general cases, the measured signal pattern may look like that given in Fig. 3.1.

Figure 3.2 gives the probability distribution of deviation of the measured signal with respect to an ideal sine that corresponds to the probe signal. We will call this sine a "general curve". The possibly random character of dots in the plot (Fig. 3.2) does not mean that the information cannot be analyzed statistically. Some of the points may be excluded from consideration once a limit on the deviation is specified. Assuming a tolerance of 7 dB as a limit, we can build a bar diagram of the distribution as shown in Fig. 3.3.

Now, as we continue with the graphical interpretation of the data, it becomes clearer that the distribution that we deemed to be random at the beginning is not random in deed. The next step is to build a curve over the maximum values of the bars (see Fig. 3.4).

In Fig. 3.4, we see that the tendency is clearly defined, something that was not obvious in the deviation distribution in Fig. 3.2. This trend determination is the first step in input signal processing. The determination case is simplified, however.

The second step is to determine whether the leftover random variation is truly random. In a number of cases, it is not, in which case, the

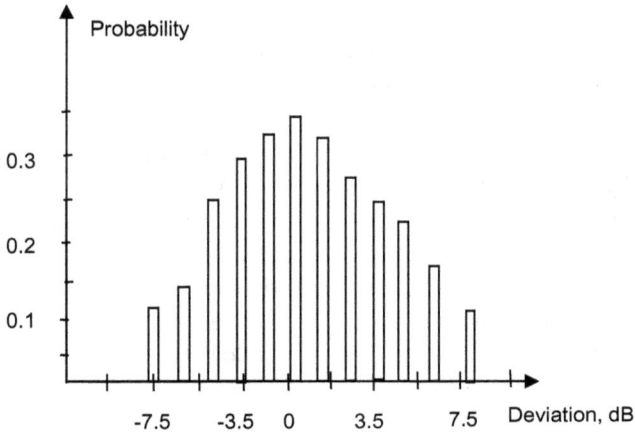

Fig. 3.3.   Bar-diagram of probability of deviation from the general curve.

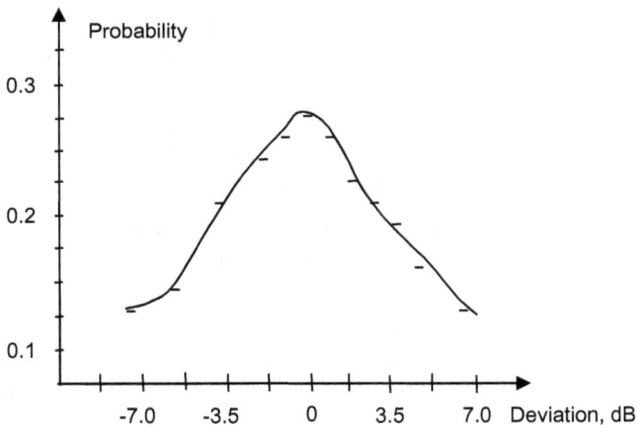

Fig. 3.4.   Same as Fig. 3.3 but approximated with a curve.

deterministic part may be singled out again and recorded into the system's memory. The random part's range is then compared to the pre-installed tolerance range. If the random part falls within the tolerance range, the disturbance (i.e. the input signal) is identified on the basis of the examples in the computer memory.

The identification of the above (leftover) part takes place by fitting simple models to localized subsets of the data (the same leftover parts) to build up a function that describes the deterministic part of the variation in data, point by point. In other words, we do not have to specify a global function of any form to fit a model of the data, only to fit segments of the data. In that sense, the word "deterministic" may not be truly correct since only segments are processed. The method used here is similar to the locally weighted polynomial regression proposed by Cleveland in 1979 and further developed by Cleveland and Delvin in 1988.

The principle is to give more weight to the points nearest to the point of estimation and the least weight to the points that are farthest away. The premise is that the point nearest to the estimation point and the estimation point itself are more likely to be related to each other. Following this logic, the points that are likely to follow the local model have the greatest influence on the local model's parameters' estimates. Rephrasing this principle, we can state that the points that are less likely to conform to the local model, have less influence on the local model's parameters' estimates.

In the process of the model development, the analyst provides a smoothing parameter value and the degree of the local polynomial. The local polynomials that fit each subset of data are almost always of first and second degrees, i.e. either locally linear (straight line) or locally quadratic. Using a zero degree polynomial turns the model function into a weighted moving average. The idea of the discussed approach is that any function can be well approximated in a small neighborhood by a low order polynomial and that simple models can be fit to data easily. High-degree polynomials would tend to overlap the range of data in each subset thus making accurate computations more difficult.

Once the model for each input signal is developed, the analyst "marks" each particular model that corresponds to the false signal. The marked models form a "window" for the detection system that allows such marked functions to pass "unnoticed" each time when the signal corresponding to such models is identified by the system.

Each current model may be modified by the analyst based on the principle described above; specifically, the analyst determines the bandwidth ("smoothing parameter") that specifies how much of the data is used to fit each local polynomial. The smoothing parameter, q, is a number between $(d+1)/n$ and 1, with d denoting the degree of the local polynomial. The value of q is the proportion of data used in each fit. The subset of data used in each weighted fit of least squares comprises the nq (rounded to the next integer) points whose explanatory variables (values) are the closest to the point at which the response is being estimated.

The flexibility of the approximation depends on the q-number: the larger the value is, the less sensitive is the fitting or vice versa. The values that are too small, however, cause the function to pick up the random error in the data. Ultimately, it is the analyst who marks the desirable and undesirable input data sequences.

Thus, the biggest advantage of the method is the fact that it does not require the specification of a function to fit the model to all of the data in the sample. Instead, the analyst–operator has to provide a smoothing parameter and the degree of the local polynomials (at the stage of the program's adjustment). Eventually, the correctness of the model is verified by the declining number of false alarms if the smoothing parameters are chosen correctly.

In addition, the method proves to be very flexible, making it ideal for modeling complex processes for which no theoretical models exist.

The main disadvantage of the above approach is the necessity to have large, densely sampled datasets in order to yield good models. This is not really surprising, however, since this approach needs extensive empirical information on the local structure of the process in order to perform the local fitting. Incidentally, this disadvantage is the flip side of the main advantage that calls for use of extensive sampling and detailed local fitting that we have been striving to implement. Another disadvantage is that the method does not produce a regression function that is readily represented by a mathematical expression. This can make it difficult to transfer the results of the input signal's analysis to an external system not possessing the same configuration and software. In order to transfer the regression function to another system, the latter would need identical software and a dataset. In nonlinear regression, on the other hand, it is only necessary to write down a functional form in order to provide estimates of the unknown parameters and the estimated uncertainty.

The easiness of adaptivity creates certain requirements for the identification to be successful. In light of this, the Nonlinear Least Squares Regression (NLSR) seems to be the most successful. The biggest advantage of the NLSR over many other techniques is the broad range of functions that can be fitted. Although many scientific and engineering processes can be described well using linear models, or other relatively simple types of models, object identification is inherently nonlinear. No matter how well we describe the features of a certain object, substance, etc. there is always something out of ordinary or nonstandard that hinders the normal procedure. However, statistically, the irregularities may be excluded or taken into account. The NLSR can produce good estimates of the unknown parameters in a model with relatively small datasets.

Another advantage is that nonlinear least squares techniques are fairly well-developed as a computational means with high computing confidence, prediction and intervals of calibration to answer scientific and engineering questions. In most cases, the probabilistic interpretation of the intervals produced by nonlinear regression is only approximately correct, but these intervals still work very well in practice.

The major cost of moving to the NLSR from simpler modeling techniques such as linear least squares is the need to use iterative optimization procedures to compute the parameter estimates. With functions that are linear, the parameters can always be obtained analytically, while that is generally not the case with nonlinear models. The use of iterative procedures requires the user to provide the starting values for the unknown parameters before the software is applied for optimization. The starting values must be reasonably close to the yet unknown parameter estimates or the optimization procedure may not converge. In our case, the library of preset values gives the initial data for processing. Thus, as already mentioned, the identification system needs calibration by a series of typical disturbances.

3.1.2. *Nonlinear Least Squares Regression (NLSR)'s definition*

NLSR extends linear least squares regression for use with a much larger and more general class of functions. Almost any function that can be written in a closed form can be incorporated in a nonlinear regression model. Unlike linear regression, there are very few limitations on the way parameters can be used in the functional part of a nonlinear regression model. The way in which the unknown parameters in the function are estimated, however, is in principle the same as it is in Linear Least Square Regression (LLSR) (Hiebert, 1981; Fox, 2002).

As the name suggests, a nonlinear model is any model of the basic form: $y = \mathrm{f}(\vec{x}, \vec{\beta}) + \varepsilon$, in which:

(1) The functional part of the model is *not linear* with respect to the unknown parameters: $\beta_0, \beta_1 \ldots$. (the functional part refers to the algebraic form of a relationship between a dependent variable and regressors or explanatory variables); x is a variable; $\varepsilon =$ random error, assumed to be normally distributed, independently of the errors for

other observations, with expectation 0 and constant variance: $\varepsilon_i \sim NID(0, \delta^2)$.

(2) The *method of the least squares* is used to estimate the values of the unknown parameters;
(3) The function is smooth with respect to the unknown parameters;
(4) The least squares criterion that is used to obtain the parameters has a unique solution.

The last two criteria are not essential parts of the definition of a nonlinear least squares model, but have a practical value. The following functions may serve as simple example of common nonlinear functions used in model building:

$$f(x; \vec{\beta}) = \frac{\beta_0 + \beta_1 x}{1 + \beta_2 x}, \tag{3.15}$$

$$f(x; \vec{\beta}) = \beta_1 x^{\beta_2}, \tag{3.16}$$

$$f(x; \vec{\beta}) = \beta_0 + \beta_1 \exp(-\beta_2 x), \tag{3.17}$$

$$f(x; \vec{\beta}) = \beta_1 \sin(\beta_2 + \beta_3 x_1) + \beta_2 \cos(\beta_5 + \beta_6 x_2). \tag{3.18}$$

3.2. Developing a model: Example

The examples given in Figs. 3.1–3.4 suggest a sinusoidal model as an appropriate one. The basic sinusoidal model is:

$$Y_i = C + \alpha \sin(2\pi \omega T_i + \varphi) + E_i, \tag{3.19}$$

where C is a constant defining a mean level, α is amplitude for the sine function, ω is the frequency, T_i is a time variable, E_i is an unobserved scalar random value and φ is the phase. This sinusoidal model can be fitted using NLSR.

To obtain a good fit, sinusoidal models require good starting values for C, the amplitude, and the frequency. A starting value that may be considered good in this case can be obtained by calculating the mean of the data. If the data show a trend, i.e. the assumption of constant location is violated (e.g. as in case of a moving object targeted by the identification system), we can replace C with a linear or quadratic least square fit. That is when the model becomes:

$$Y_i = (\beta_0 + \beta_1 T_i) + \alpha \sin(2\pi \omega T_i + \varphi) + E_i \tag{3.20}$$

or

$$Y_i = (\beta_0 + \beta_1 T_i + \beta_2 T_i^2) + \alpha \sin(2\pi\omega T_i + \varphi) + E_i. \qquad (3.21)$$

Assuming that our data did not have any meaningful change of location (as in the case with crossing the surveillance beam), we can fit the simpler model with C equal to the mean. The starting value for the frequency can be obtained from a spectral plot (in our case, Fig. 3.1 may serve as such) using, e.g. the Fourier transform of the data in Fig. 3.1. The complex demodulation phase plot (not shown) can be used to refine the initial estimate for the frequency.

In order to obtain a good starting value for the amplitude, a complex demodulation amplitude plot can be used. In essence, we are refining the initial estimate for the frequency. In addition, this plot indicates whether or not the amplitude is constant over the entire range of the data or if it varies. If the plot is essentially flat (zero slope), then it is reasonable to assume constant amplitude in the nonlinear model. However, if the slope varies over the range of the plot, we may need to adjust the model to be:

$$Y_i = C + (\beta_0 + \beta_1 T_i) \sin(2\pi\omega T_i + \varphi) + E_i. \qquad (3.22)$$

In (3.22), we replace α (from 3.21) with a function of time. A linear fit is specified in (3.22), however, in a more complex case, a more sophisticated function may be used. An example of a complex demodulation amplitude plot is given in Fig. 3.5.

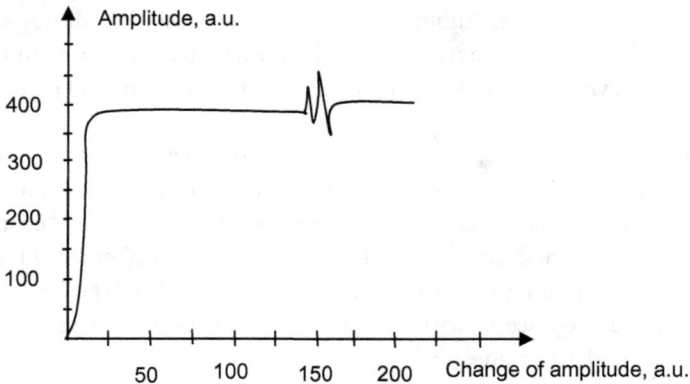

Fig. 3.5.   Complex amplitude plot.

From Fig. 3.5, we can see that:

(1) Amplitude is almost fixed at 390;
(2) There is a change in amplitude at approximately 170;
(3) There is a start-up effect.

Considering a nonlinear model, the plot shows that α may be a constant to be adequate to the given dataset.

3.3. Validation of the created model

Once the model is created, we can evaluate the fit. The first step in doing so would be the generation of a set of graphs that show the correctness of the initial assumptions. The graphs are used here for the purpose of illustration: they are not generated routinely as part of the system's operation. However, the graphs are helpful for the analyst in the process of pattern recognition development.

Figures 3.6–3.9 show the process of input data analysis. The adjustments are made automatically but the plots presented can be instrumental in the process of the system's development.

The interpretations from the figures are as follows:

1. The *Run sequence plot residuals* (Fig. 3.6) show that the data do not have any significant change of location, although there seems to be some shifts at several spots. A start-up effect mentioned earlier (Fig. 3.5) is

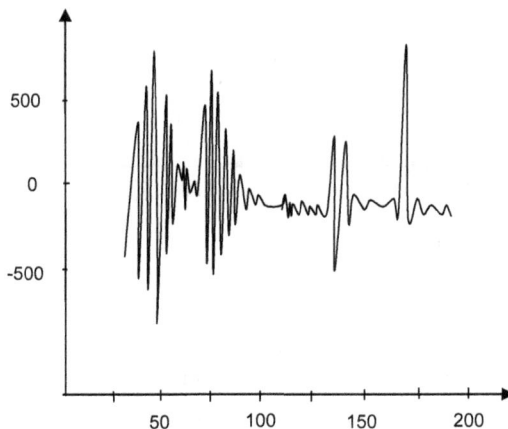

Fig. 3.6.   Run sequence plot residuals (a.u.).

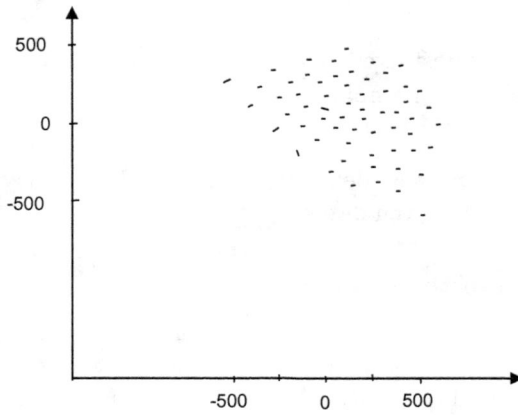

Fig. 3.7.   Lag plot residuals (a.u.).

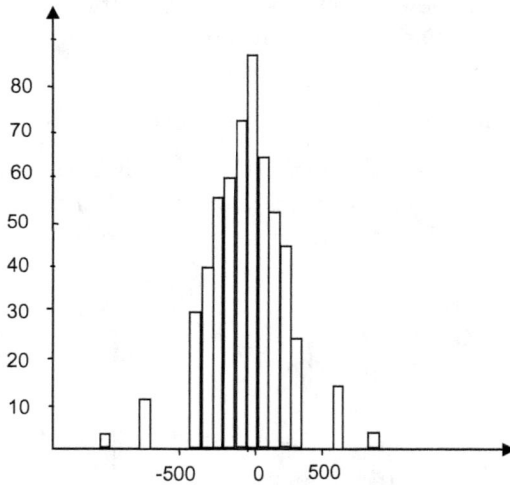

Fig. 3.8.   Histogram residuals (a.u.).

present too. Some outliers are also present. Residual of a sample is the difference between the sample and either the observed sample mean or the regressed (fitted) function value. The fitted function value is the value predicted by the statistical model.

2. The *Lag plot* shows that the data are random by appearance.

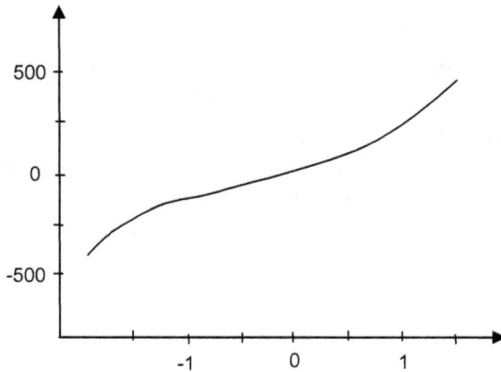

Fig. 3.9.   Normal probability plot residuals (a.u.).

3. The *Histogram* and the *Normal Probability* plot do not demonstrate serious deviations from normality in the residuals. The left portion of the *Normal Probability* curve shows some abnormality.

Thus, Figs. 3.6–3.9 provide a visual characterization of the fit that has been achieved.

The next step is to remove the outlier and improve the model. The program calculates the convergence measure, residual standard deviation, residual degrees of freedom and final parameter estimates.

Given below are the specific numbers for a given example where Yi was calculated from the Eqs. (3.19)–(3.22).

Example:

The original fit with the residual standard deviation of 155.84 was 155.84. Plugging in the known numbers:

$$Yi = -178.79 - 361.77(2\pi \times 0.302596T_i + 1.465) + E_i.$$

The new fit with a residual standard deviation of 148.34 is:

$$Yi = -178.79 - 361.76(2\pi \times 0.302597T_i + 1.465) + E_i.$$

There is a small change in the parameter estimates and about a 5% reduction in the residual standard deviation. In this case, removing the residuals, we hardly reduce the model's variability. We see that the frequency ω gave a little improvement in adjustment. Although real life calculations may be more extensive and less precise, the example illustrates the principle of calculations reasonably well.

3.4. Spectrum analysis

One of the main components of target identification is spectrum analysis. In general, the input signal has a complex spectrum. Decomposition of such a spectrum is important since once we know what the spectrum is, we can model it. Modeling is the basis for identification (as was described in the previous sections.) Spectrum analysis is not always necessary to perform and may be omitted when possible in order not to make the identification system too complicated. On the other hand, spectrum analysis is indispensable for anything that lies outside the straightforward comparison. That is why, such an analysis is considered separately from model-making.

3.4.1. *General structure model*

The purpose of spectrum analysis is to decompose the original series into underlying sine and cosine functions of different frequencies in order to determine those that appear particularly strong or important. One way to do it would be to present a spectrum as a linear Multiple Regression problem, where the dependent variable in the observed time series and the independent variables are sine functions of all possible (discrete) frequencies. Such a linear multiple regression model may be written as:

$$x_t = a_0 + \Sigma[a_k \times \cos(\lambda_k \times t) + b_k \times \sin(\lambda_k \times t)], \quad (\text{for } k = 1 \text{ to } q).$$
$$(3.23)$$

Following the commonly accepted notation from classical harmonic analysis, in this equation λ is the frequency expressed in terms of radians per unit time, that is $\lambda = 2\pi\nu_k$, where $\nu_k = k/q$. It is important to point out that the computational problem of fitting sine and cosine functions of different lengths to the data can be considered in terms of a multiple linear regression. Note that the cosine parameters a_k and sine parameters b_k are regression coefficients that tell us the degree to which the respective functions are correlated with the data. On a whole, there are q different sine and cosine functions. However, we cannot have more sine and cosine functions than there are data points in the series. If there are N data points in the series, then there will be $N/(2+1)$ cosine functions and $N/(2-1)$ sine functions. Thus, spectrum analysis will identify the correlation of sine and cosine functions of different frequency with the observed data. If a large correlation (sine or cosine coefficient) is identified, one can conclude that there is a strong periodicity of the respective frequency in the data.

Example:

To illustrate spectrum analysis' applications, let us create a series with 16 cases following the Eq. (3.23), and then we can "extract" the information from it.

First, let us create a variable and define it as:

$$x = \cos(2\pi 0.0625(\nu_0 - 1)) + 0.75\sin(2\pi 0.2(\nu_0 - 1)).$$

This variable is made up of two underlying periodicities: the first at the frequency of $\nu = 0.0625$ (or period $1/\nu = 16$; one observation computes $1/16$th of a full cycle, and a full cycle is completed in 16 observations) and the second at the frequency of $\nu = 0.2$ (or period of 5). The spectrum analysis summary is shown in Table 3.1.

Let us now review the columns. Clearly, the largest cosine coefficient can be found for the 0.0625 frequency. A smaller sine coefficient can be found at frequency ~ 0.1875. Thus, the two sine/cosine frequencies which were "inserted" into the example data file are reflected in Table 3.1. The insertion implies two components that exist. A (signal from the target) beam disturbance from an object may be decomposed into series of insertions, each one of which represents a certain feature of the object. The sine and cosine functions are mutually independent (or orthogonal); thus, we may sum the squared coefficients for each frequency to obtain the periodogram. Specifically, the periodogram values above are computed as:

$$P_k = sine\ coefficient_k^2 + cosine\ coefficient_k^2\ N/2,$$

where P_k is the periodogram value at frequency ν_k and N is the overall length of the series. In particular, the periodogram values can be interpreted

Table 3.1.

Spectral analysis. No of cases: 16

t	Frequency	Period	Cos coef.	Sine coef.	Periodogram
0	0		0	0	0
1	0.0625	16	1.006	0.028	8.075
2	0.125	8	0.033	0.079	0.059
3	0.1875	5.33	0.374	0.559	3.617
4	0.25	4	−0.144	−0.144	0.333
5	0.3125	3.2	−0.089	−0.06	0.092
6	0.375	2.67	−0.075	−0.031	0.053
7	0.4375	2.29	−0.07	−0.014	0.04
8	0.5	2	−0.068	0	0.037

in terms of the variance (sums of squares) of the data at the respective frequency or period. Periodogram, therefore, represents the sum of the object features as far as its identification is concerned.

3.5. Signal processing implementation

3.5.1. *General description*

The output signal from the receiver is amplified and transformed into a digital form. Since the output of the receiver is arbitrary in principle (in practice, however, it is not so in most cases), it is compared with the library of possible outputs. Pattern recognition is the basis for comparison. Recognition is simulated in two stages: first, a model is built based on the receiver's output information (Fig. 3.10). Then, the model is compared to the signal. The comparison between the model and the original signal gives the degree of correlation between the two (the signal processing is digital). It is not necessary for the sequence of the signal code to coincide with the library patterns completely since Automatic Pattern Recognition Algorithm (APRA) calculates the approximation that should correspond to one of the library patterns with a predetermined tolerance. Once the intrusion signal's type is identified, the system blocks the input (e.g. the one that corresponds to a typical obstacle, such as precipitation) or gives an alarm signal (Sokolnikov, 2006b).

3.5.2. *Process simulation and identification*

The digitized signal from the input initiates its simulation as shown in Fig. 3.10.

The basis for simulation is provided by the library of possible target signal. In the following example, we have library models of possible signals.

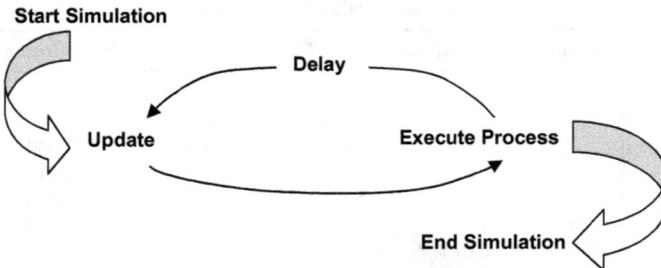

Fig. 3.10.   Process simulation sequence.

In our case, we consider two signals: a and b. In Step 1, a signal from the target is received. Step 2 gives a command on comparison of the signals and identification of possible matches. Step 3 presents the comparison itself. And in Step 4, the decision is made whether both (a and b) or either of them matches any signals in the library.

Example:

(1) Library MSSS;
(2) Use MSSS*std_logic_1164.all
(3) Entity example_1 IS
      PORT (a,b:IN std_logic; z:OUT
  Std_logic);
(4) ARCHITECTURE type 1 OF example_1 IS
                $z \leq$ (a AND (NOTb))OR((NOTa)ANDb);
(5) END type1;

Process features:

1. All processes are executed concurrently;
2. Concurrent signal assignments-statements are actually one-line processes;
3. Processes are re-executed if any signal in its sensitivity list is changed;
4. Statements are executed sequentially within a process;
5. Concurrent processes with sequential execution within the process offer maximum flexibility;
6. Various levels of abstraction are supported;
7. Modeling of concurrent and sequential events is observed in real systems.

The order of disturbance signals may be different; consequently, it enforces the requirement of concurrent signal assignments and statements to be executed in parallel. The following example is the same as the previous one except that Step 6 and the step following it show that both a and b signals are processed in parallel taking into account their delays. This feature allows for different signals to be included in one model creation. Thus, we can observe several events of identification at the same time with a subsequent decision that takes into account all coming disturbances.

Example:

(1) Library MSSS
(2) Use MSSS, std_logic_1164.all

(3) ENTITY example_1 ls
        PORT (a,b: IN std_logic;z: OUT std_logic);
(4) END example_1;
(5) ARCHITECTURE type 2 OF example_1 ls
(6) SIGNAL s1, s2, s3, s4: std_logic: = '0';

BEGIN
(7) s1 <= NOT b AFTER 3 NS;
(8) s4 <= b AND s2 AFTER 5 NS;
(9) z <= s3 OR s4 AFTER 5 NS;
(10) s2 <= NOT a AFTER 3 NS;
(11) s3 <= a AND s1 AFTER 5 NS;
(12) END type2;

3.6. Conclusions

The suggested computational method based on TSA was given to illustrate the principles with which to approach the problem of the implementation of a mobile THz system that allows automatic or semi-automatic calculations and the display of identification parameters. It would be impossible to present a complete portfolio of an even, single computational process from beginning to end. However, it was worthwhile to indicate a number of major steps, procedures and features pertaining to the above process. The author's goal was also to show how to approach the data analysis and what possible outcome we may receive. At the same time, I hope that the provided references may be an additional source of information for those readers who are willing to broaden their understanding of mathematical methods that are employed, or that may be employed for THz identification. In addition, this chapter's objective was to point out the other areas of technological and scientific expertise where the reader may find collaborators to work with on a specific project.

References

Backstrom, T and P Alku (2003). All-pole modeling technique based on weighted sum of LSP polynomials. *IEEE Signal Processing Letters*, 10(6), 665–668.

Chatfield, C (1989). *The Analysis of Time Series (An Introduction)*, 4th edn. Chapman and Hall.

Fox, J (2002). Nonlinear regression and nonlinear least square. *Companion to Applied Regression*, Appendix to an R and S-plus companion to applied regression. Sage Publications.

Gonzalez, R and R Woods (1992). *Digital Image Processing*, pp. 187–213. Addison Wesley.

Hamilton, JD (1994). *Time Series Analysis*. Princeton University Press.

Hiebert (1981). An evaluation of mathematical software that solves nonlinear square problems. *ACM Transactions on Mathematical Software*, 7(1), 1–16.

Jain, A (1989). *Fundamentals of Digital Image Processing*, pp. 244–253, 273–275. Prentice Hall.

Kellert, SH (1993). *In the Wake of Chaos: Unpredictable Order in Dynamical Systems*. University Of Chicago Press.

Marion, A (1999). *An Introduction to Image Processing*. Chapman and Hall.

Mills, TC (1990). *Time Series Techniques for Economists*. Cambridge University Press.

Pandit, SM and SK Wu (1983). Ming *Time Series and System Analysis with Applications*. John Wiley & Sons, Inc.

Percival, DB and AT Walden (1993). *Spectral Analysis for Physical Applications*. Cambridge University Press.

Percival, DB and AT Walden (2000). *Wavelets Methods for Time Series Analysis*. Cambridge University Press.

Sokolnikov, A (2006a). Adaptive non-intrusive terahertz identification. *Proc. SPIE*, 6212.

Sokolnikov, A (2006b). Mobile security surveillance system. *Proc. SPIE* 6201.

Chapter 4

Physics of Producing and Detecting THz Waves

The first THz sources were thermal and produced low-power incoherent radiation. The alternative options, such as free-electron lasers or optically-pumped gas lasers were highly complex and bulky. Far-infrared Fourier spectroscopy was widely used for experiments at the frequencies that now correspond to the higher end of the THz range. Liquid helium-cooled bolometers, which deliver a signal proportional to radiation intensity, with a relatively poor noise performance, were traditionally used as the most sensitive detectors. Recent advances in the fields of ultrashort pulsed lasers, nonlinear optics and crystal growth techniques have given commercially available sources of bright, coherent, broadband THz pulses and enabled room temperature detection. Well-established THz-TDS (Time Domain Spectroscopy) is one of the methods that makes it possible to record not just intensity, but a time-resolved amplitude of the electric field.

Typically, the THz pulses are generated by an ultrashort-pulsed laser and last only a few picoseconds (ps). A single pulse can contain frequency components covering the whole THz range — from 0.05 to 10 THz. For detection, the electrical field of the THz pulse is sampled and digitized similar to the way an audio card transforms electrical voltage levels in an audio signal into numbers that describe the audio waveform. For example, in THz-TDS, the electrical field of the THz pulse in the detector interacts with a much shorter laser pulse (0.1 ps is a typical number) which results in an electrical signal that is proportional to the electric field of the THz pulse at the time the laser pulse switches (gates) on the detector. This procedure is repeated by varying the timing of the laser pulse gating in order to scan the THz pulse and construct its electric field as a function of time. Subsequently, the received data is processed (Fourier transform is used) to extract the frequency spectrum from the time-domain data.

Other sources of THz radiation include *multipliers*, such as doublers and triplers. Schottky diode frequency multipliers and multiplier chains are currently capable of producing THz radiation in the range of about 100 GHz–2.5 THz. The efficiency across this broad range of frequency and performance can be well-described by a simple exponential decay model. This model can also be used to predict achievable performance for Schottky diode frequency multipliers and multiplier chains. However, the output power remains on the order of 1 μW or even lower and often requires cryogenic temperatures (Ward *et al.*, 2006).

In the time domain, *photoconductive sampling* can be used for THz generation. The method has been used by many groups for THz spectroscopy in free space and on transmission lines. These systems consist of two fast photoconductive switches that are excited by a mode-locked laser and are coupled to each other via antennas and a transmission line. In general, photoconductive sampling is a technique of optical sampling, based on the use of photoconductive switches. On such a switch, a short laser pulse can close an electrical connection for a very short time and produce a pulse with duration in the order of picoseconds. For example, in the frequency domain, photomixers that use low-temperature-grown GaAs (LTG-GaAs) photoconductors illuminated by two continuous-wave (CW) diode lasers have generated CW difference-frequency radiation from a few MHz to 5 THz. These sources have applications such as broadly tunable local oscillators for submillimeter-wave receivers and systems for high-resolution gas spectroscopy when coupled to a cryogenic detector such as a bolometer (Verghese *et al.*, 1998).

4.1. Interaction of THz radiation with matter

Interaction of THz waves with different materials and substances has a more specific character than that in electromagnetic regions that have been used for matter identification so far. Actually, these specific features make possible a more detailed characterization of different organic and inorganic subjects of interest. THz waves cannot penetrate metals and water, although nonpolar liquids are reasonably transparent to THz radiation. Many dry materials such as plastics, paper, cloth, etc. are transparent to THz rays, though saturating them with water greatly impairs their transparency. Many gases reveal sharp resonances in the THz region, as do a number of explosives. Characteristic "finger-prints" of the THz radiation make the identification of objects highly attractive for military and special purposes, as well as for medical industrial and other applications. Below is a more detailed description of the physical mechanism of THz interaction with matter.

The interaction of THz radiation of frequency ν with an object can be expressed in terms of the frequency-dependent complex refractive index $N(\nu)$ of the material:

$$N(\nu) = n(\nu) + i\frac{c\alpha(\nu)}{4\pi\nu}, \tag{4.1}$$

where $n(\nu)$ represents the real refractive index as a function of the frequency ν, c is the speed of light in a vacuum and $\alpha(\nu)$ is the absorption coefficient. THz-TDS makes it possible to determine an object's full complex refractive index by coherent detection of the amplitude of the electric field of the THz pulse.

Complex refractive indices are usually determined by THz-TDS by comparison of the reflected or transmitted fields propagated within the object of interest, with the incident reference pulse E_0. If transmitted field is of amplitude E_T at location r, and frequency ν propagates through a slab of thickness d, the complex transmittance coefficient, t, is given by (Born and Wolf, 1959):

$$t = \frac{\bar{E}_T(r,\nu)}{\bar{E}_0(r,\nu)} \approx t_{01}(\bar{r},\nu)t_{02}(\bar{r},\nu) \exp\left[i\frac{2\pi\nu}{c}n(\bar{r},\nu)d\right] \exp\left[-\frac{\alpha(\bar{r},\nu)}{2}d\right], \tag{4.2}$$

where we retain in-plane r-dependence of the values describing the differences of the optical properties in the plane perpendicular to the propagation of light.

Transmission fields $t_{01}(\bar{r},\nu)$ and $t_{02}(\bar{r},\nu)$ are the Fresnel coefficients for the front and back surfaces of the slab respectively. From equations (4.1) and (4.2) we can receive the absorption $\alpha(\bar{r},\nu)$, and refractive $n(\bar{r},\nu)$ coefficients as well as spectral dependence for local interfaces. The measurement of the above coefficients is possible with the use of coherent methods of the THz spectrum acquisition. A similar approach can be used to determine these coefficients from reflection measurements.

Coherent methods (such as THz-TDS) revealed characteristic resonances in gases, liquids and solutions in the THz range, corresponding to rotational and vibrational energy modes of the molecules, and broader absorption peaks in solids and liquids. Other substances that have been characterized in the THz region of the spectrum include semiconductor materials and dielectrics.

Spectroscopic methods are also used in the biological sciences for investigating DNA, protein and tissues. The characteristic spectra *in vivo* depend not only on the constituent materials but also on the thermal or physical arrangement of the material's molecules. This means that *in vivo* measurements on tissues will be different from spectroscopic measurements on homogeneous samples of molecules or proteins.

Many nonpolar liquids are reasonably transparent to THz radiation, while polar liquids such as water and alcohols strongly attenuate THz radiation. Dry materials such as plastics, paper, cardboard, cloth, etc are readily transparent. As it is shown in Fig. 4.1, when saturated with water dry materials, generally transparent to THz waves, can become highly absorbing.

Measurements of the absorption coefficient of water show that it strongly attenuates signals in the THz region, with broad absorption peaks at 6 THz and 19 THz (Kindt and Schmuttenmaer, 1996). This has strong implications for the use of TPI for medical imaging and forensics. On the one hand, the high absorption limits the depth of penetration of signals through

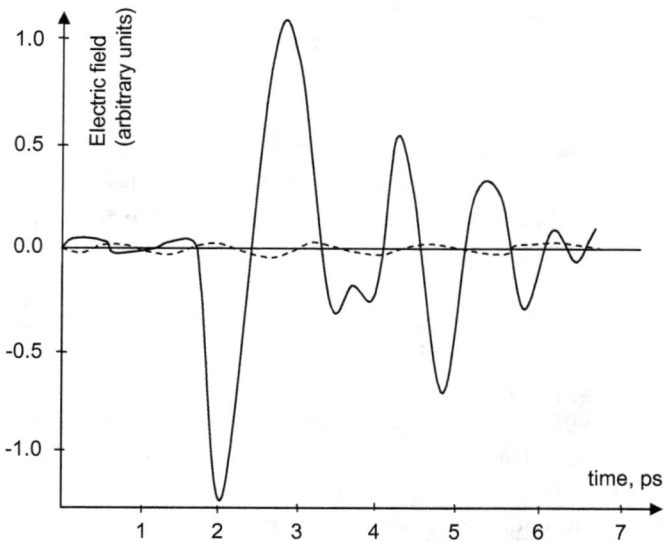

Fig. 4.1.   THz pulse transmitted through 3 mm of dry cotton cloth (solid line) and the same material saturated with water (dotted line).

tissues with high moisture content, on the other, it makes THz methods very sensitive to changes in the water content of tissue. Measurements by Bezant (2000) demonstrated that at the present moment THz probing signal may be detected after penetration through at least 1.5 mm of moist skin tissue with the detection signal-to-noise ratio (SNR) is 500 to 1. This is considerably further than through water alone. Mittleman *et al.* (1996) addressed the question of sensitivity to water content and Suggested that for an SNR of 100, a minimum concentration given by $n \times x = 10^{16}$ cm^{-2} can be detected, where n is the density of water molecules and x is the path length.

4.1.1. *Photoconductive THz generation: Photoconductive emitters*

Photoconductive (PC) emitter is a semiconductor device, in which an optical laser pulse (100 fs or shorter) creates carriers (electron–hole pairs), thus making the semiconductor a conductor. This results in a sudden electrical current across a biased antenna pattern on the semiconductor. This changing current causes the emission of THz waves, which is similar to the process of producing of radio waves in a radio transmitter's antenna (Fig. 4.2).

Usually, the antenna has two electrodes that are patterned on a low-temperature GaAs, semi-insulating GaAs or other semiconductor (*e.g.* InP)

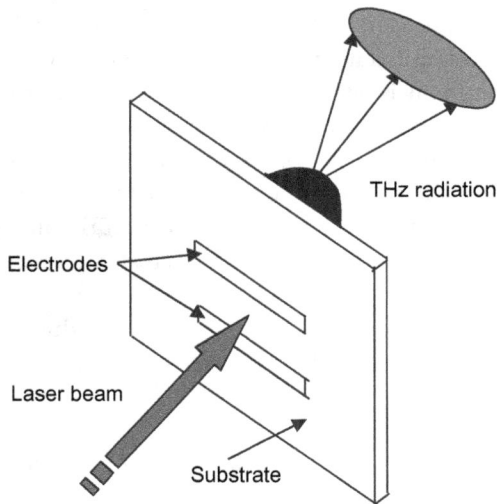

Fig. 4.2.   Principle of functioning of photoconductive emitter.

substrate. Typically, the electrodes form a simple dipole antenna with a gap (approximately, several micrometers wide). The electrodes are biased with voltage of several dozens (30–40 kV) kV to excite electrons across the band gap of the semiconductor substrate. Presently, the duration is 100 fs or less to produce the above effect. Generally, the photon energy of the laser excitation pulse should be higher than the band gap of the substrate. Sometimes, the excitation pulse with lower photon energy can be used to generate free carriers that produce photocurrent being driven by the bias applied to the electrodes. Since electrons usually have much higher mobility than holes, the mobility of the latter may be ignored. The current density then is:

$$J(t) = N(t)e\mu_e E_b, \tag{4.3}$$

where $N(t)$ is the density of photocarriers, e is the electron charge, μ_e is electron mobility, E_b is the electric field bias.

The energy of the THz pulse comes from the electric energy across the gap between the electrodes rather than from the laser pulse energy. Thus, the pulse energy of THz waves is not limited to the pulse energy of the laser excitation beam and conversion efficiency (from optical laser to THz pulse energy) greater than one is possible. The excitation acts more like a trigger releasing energy stored between the electrodes on the substrate into THz waves. Under weak excitation, the THz wave's energy is proportional to the laser excitation energy. On the other hand, increasing the bias field also has a limit, since a high electric field may cause a dielectric breakdown in the semiconductor substrate. The nature of the breakdown is the same as it is in all semiconductors: it is either electrical or thermal. Most breakdowns of PC emitters are thermal, although some may be electrical or field-induced breakdowns for very high biases or narrow gaps between electrodes. Thus, we are restricted in the amount of energy that we can accumulate between the electrodes in order to produce high intensity THz pulses. Better results may be achieved by proper semiconductor substrate fabrication or special coating.

Since THz radiation detected through photoconduction is caused by a flow of charge, it is described by Maxwell's equations. In the differential form:

$$\nabla x E = -\frac{\delta B}{\delta t} \qquad \text{(Faraday's law of induction)}, \tag{4.4a}$$

$$\nabla x B = \mu_0 J + \mu_0 \varepsilon_0 \frac{\delta E}{\delta t} \qquad \text{(Ampere's circuital law)}. \tag{4.4b}$$

Here, B is magnetic field; J is total current density and E is electric field. In general, any process that creates a time-dependent change in the material properties μ or ε can act as a source term of emission of THz radiation. In particular, for THz generation using photoconductive switches, an ultrashort-optical pulse incident on a semiconductor causes rapid transient changes to the macroscopic material properties represented by charge density $\rho(t)$, permeability $\mu(t)$, and permittivity $\varepsilon(t)$. The major change caused by the optical pulse is assumed to be in the conductivity, σ. The rapid, optically-induced charge, ρ on a femtosecond time scale is the origin of ultrafast THz pulses generated through photoconduction as well as the physical mechanism by which the THz pulses are detected. In addition, there are many physical processes that occur when ultrafast-optical pulses are absorbed by semiconductors, and, from an experimental viewpoint, they are not easily separable. Both resonant (absorption of a photon to create charge carriers) and nonresonant (nonlinear optical difference frequency generation) effects can contribute to THz pulse generation. As an example, we can take GaAs, in which both processes take part in TH pulse generation. The driving term in Maxwell's equations, $\rho(t)$, results from an ultrafast optical pulse. The resulting electric and magnetic fields created through the time-dependent conductivity have fast transients with correspondingly broad spectral bandwidths. The spectral bandwidth can be estimated from the uncertainty relation (between the bandwidth, $\Delta\omega$ required to support a transient signal, and Δt).

Since $\Delta t \Delta\omega \geq 1/2$, a ps excitation pulse can create spectral widths in the order of 0.3 THz whereas a 100-fs transient has a bandwidth of at least 3 THz. With sub-10 fs pulses already commercially available from ultrafast lasers, a very broad spectral THz range is available. An example of a THz measurement is shown in Fig. 4.3. Figure 4.3(a) compares the measurement of the electric field of the 50-fs optical pulse at $\lambda = 800$ nm used to generate the THz pulse. The shape appears to be a Gaussian envelope. The spectral amplitude of the THz pulse and optical pulse are given in Fig. 4.3(b). One of the advantages of ultrafast THz measurements is that the bandwidth allows a broad range of energy frequencies to be probed in a single measurement with high temporal resolution.

Figure 4.3 highlights some of the differences between short pulses in the THz and optical spectral regions. The electric field of the THz pulse is a nearly single cycle oscillation. The optical pulse here is described as a Gaussian pulse envelope superimposed on a sinusoidal carrier frequency. The optical pulse satisfies the slowly varying envelope approximations,

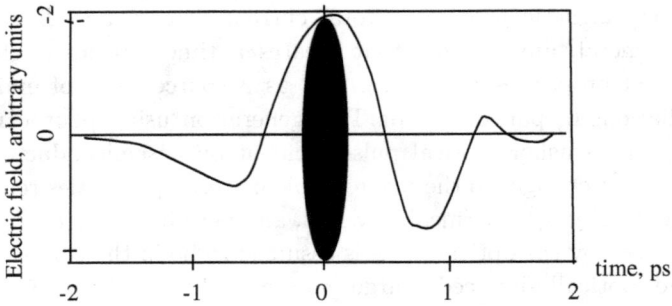

Fig. 4.3a.   (a) Electric field of a measured THz pulse and the stimulated field of the optical-pulse excitation are shown in this figure. The amplitude spectrum is shown in (b).

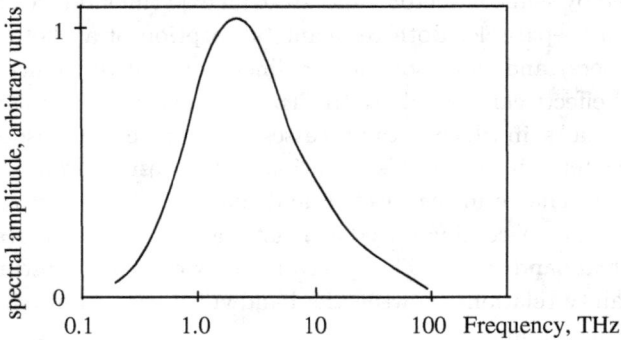

Fig. 4.3b.   (b) Electric field of a measured THz pulse and the stimulated field of the optical-pulse excitation are shown in (a). In this figure the amplitude spectrum is shown.

whereas the THz pulse does not. Thus, many of the simplifying assumptions used in quantum nonlinear optics do not necessarily apply to the THz systems. Theoretically, an accurate description of the pulse propagation requires the solution of the coupled Maxwell–Bloch equations. Since THz pulses used for a number of applications (*e.g.* spectroscopy) have very low peak powers compared with optical pulses, nonlinearities can generally be neglected and propagation is described using linear dispersion theory.

   One of the most significant difference between THz pulses and visible, infrared, or ultraviolet light is that THz spectroscopy directly measures the electric field of the pulse rather than the intensity. It is very important feature (along with some other ones) that makes possible THz identification.

   The direct measurement of field is possible because of the much slower variation of the field (Fig. 4.3). Intensity, I and electric field, E are related by

$$I = 1/2\sqrt{\frac{\mu E E^*}{\varepsilon}}. \tag{4.5}$$

Equation (4.5) illustrates the well-known fact that intensity measurements contain no phase information unless techniques such as holography or interferometry are used to measure phase relative to another optical beam. However, the amplitude and phase of the time-resolved electric field of the THz pulse can be detected coherently, directly from THz measurements. For the more familiar intensity measurements, energy conservation dictates that the reflection and transmission coefficients are real and positive with a magnitude less than one in almost all commonly encountered cases. Since THz experiments measure the electric field amplitude, the Fresnel coefficients can be complex and have values less than zero or greater than one. For example, the amplitude transmission coefficient from air to silicon is $t_{12} = 0.45$, whereas that from silicon to air is $t_{21} = 1.55$.

The final commonly encountered difference between experimental measurements at optical and THz frequencies is the effect of phase shifts on the measured pulse shape. Larger frequency-dependent phase shifts, such as those caused by dispersive media, do result in pulse broadening. In contrast, small-frequency, independent-phase changes result in significant reshaping of THz pulses, as shown in Fig. 4.4. Numeric phase changes of $0, \pi/16, \pi/8, \pi/4, \pi/2$ and π result in pulse reshaping up to a complete inversion of the pulse. Such phase shifts can occur on reflection, when the pulse propagates through a caustic, or through materials with a complex index.

Below, there is an example of THz generation, which is based on an application of a titanium–sapphire laser emitting ultrashort pulses.

4.1.1.1. *Example*

In a photoconductive emitter, the optical-laser pulse (100 fs or shorter) creates carries (electron–hole pairs) in a semiconductor material (*e.g.* ZnTe, a zincblende). This conduction leads to an abrupt initiation of electrical current across a biased antenna for emitting THz waves in the process similar to emission of radio waves of a radio frequency transmitter. The ultrafast (100 fs) laser pulse must have a wavelength that is short enough to excite electrons across the bandgap of the semiconductor substrate. A typical energy level of the applicable Ti sapphire oscillator laser is about 10 nJ. The more practical values of the laser pulse of about 1 mJ would

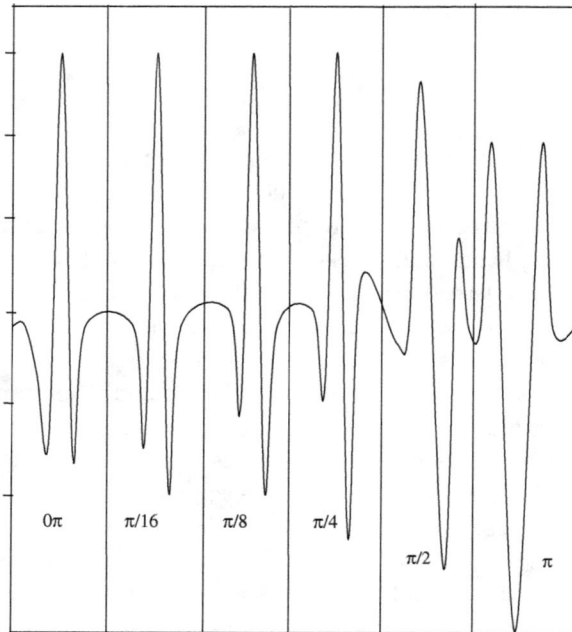

Fig. 4.4. Effect on frequency-independent phase shifts from 0 to $\pi/2$ radians on the THz pulse shape.

require a significant increase of the bias voltage in the semiconductor of up to 10 kV. This aspect of THz radiation generation creates difficulties while making portable THz devices that are often preferable for security and defense applications.

4.1.2. *Photoconductive detectors*

Physical principle of operation of a PC THz detector is similar in its operation to that of the emitter. The only important difference is that as a detector, the two electrodes on the semiconductor substrate are connected to a sensor but not to a high voltage power supply. By controlling the time delay between the THz pulse and the (optical) laser probe pulse, the electric field across the electrodes of a PC detector at any given time can be sampled by the probe pulse (the probe pulse, in this case, generates transient photocarriers in the semiconductor substrate). Since the THz pulses and probe pulses are correlated for a certain time delay, the photo-induced carriers see a steady electric field that produces a current (from the photo-induced carriers) between the two electrodes. The current is:

$$\bar{J} = \bar{N} e \mu_e E(\tau), \tag{4.6}$$

where $\bar{N}$ is average electron velocity, e is electron charge, μ_e is electron mobility and $E(\tau)$ is electric field as a function of the temporal delay.

As we can see from (4.6), the detection of THz pulses measures the electric field, rather than intensity. Thus, we measure not only the amplitude but also the phase of the THz pulse, while the latter is not the case for continuous (CW) THz measurements.

Photoconductive detection is similar to photoconductive generation. In this case, the bias electrical field across the antenna leads is generated by the electric field of the THz pulse focused onto antenna. The presence of the THz electric field generates current in the antenna, the signal from which is usually amplified by a low-bandwidth amplifier. This amplified current is the measured parameter, which corresponds to the THz field strength.

A part ("detection pulse") of the same ultra short laser pulse that was used to generate the THz pulse is fed to the detector, where it arrives simultaneously with the THz pulse. The detector produces different electrical signals depending on whether the detection pulse arrives when the electric field of the THz pulse is low or high. In addition, an optical delay line is used to vary the timing of the detection pulse.

4.2. Nonlinear optical pulse generation and detection

A significant amount of research going back to as early as the 1970's has explored the generation of THz waves by optical rectification of pico-and femtosecond optical pulses in nonlinear optical crystals. Experimental work in the early 1990's reported that THz radiation could be also generated by femtosecond optical pulses incident on semiconductor surfaces (Wilke and Sengupta, 2007).

Nonlinear optical techniques for producing the THz radiation and its detection are important since purely electronic methods for the above purpose are still technologically challenging. In addition, one of the research directions in the THz range is to increase the bandwidth of short THz pulses. In this aspect, nonlinear optical phenomena such as optical rectification and linear electro optical effect (Pockels'effect) are advantageous options for THz broadband generation and detection. Optical rectification and the linear electrooptic effect (Pockels effect) are nonlinear methods for the generation and detection of freely propagating subpicosecond THz pulses.

In general, optical rectification refers to the development of a DC or low-frequency polarization when intense laser beams propagate through a crystal. The Pockels' effect describes a change of polarization of a crystal with an applied electric field. Please note that optical rectification and the linear electro-optic effect occur only in crystals that are not centrosymmetric. However, optical rectification of laser light by centrosymmetric crystals is possible if the symmetry is broken by a strong electric field. Also, generation and detection of THz pulses by optical rectification and Pockels' effect require that the crystals are sufficiently transparent at THz (and optical) frequencies.

Freely propagating subpicosecond THz pulses are generated by optical rectification of femtosecond (fs) near-infrared laser pulses in crystals with appropriate nonlinear properties. The propagating THz pulses are received and detected by measuring the phase modulation of a fs near-infrared laser pulses propagating through an electrooptic crystal simultaneously with a THz pulse. The electric field of the THz radiation induces a phase modulation of the fs laser pulse by electro-optical effect.

In 1893, Friedrich Pockels discovered the linear electrooptic effect. Subsequently, the phase modulation of optical laser light at microwave frequencies using the linear electrooptic effect was demonstrated in 1962 by Harris *et al.* In 1982, Valdmanis and co-workers built the first electrooptic sampling system with picosecond resolution for the measurement of ultrafast electrical transient. In 1995, electro-optical sampling of freely propagating THz radiation pulses was demonstrated by Wu and Zhang and in 1996 by Jepsen *et al.* and Nahata *et al.*

The discussion of optical rectification with Pockels effect starts with considering scalar relationship of polarization P, electric susceptibility and electric field, E:

$$P = \chi(E)E. \qquad (4.7)$$

The electric polarization P of a material is proportional to the applied electric field E. Here, $\chi(E)$ is the electric susceptibility. The nonlinear optical properties of the material are described by expanding $\chi(E)$ in powers of the field E:

$$P = (\chi_1 + \chi_2 E + \chi_3 E^2 + \chi_4 E^3 + \cdots)E. \qquad (4.8)$$

Optical rectification and the linear electrooptic effect are second order nonlinear optical effects P_2^{nl} and described by the $P_2^{nl} = \chi_2 E^2$ term in the expansion.

As an example, let us consider an optical electric field E described by $E = E_0 \cos \omega t$. In this case, the second-order nonlinear polarization P_2^{nl} consists of a DC polarization $\chi_2 E_0^2/2$ and a polarization with a $\cos 2\omega t$ dependence:

$$P_2^{nl} = \chi_2 E^2 = \chi_2 \frac{E_0^2}{2}(1 + \cos 2\omega t). \qquad (4.9)$$

The DC polarization results from the rectification of the incident optical electric field by the second-order nonlinear electric susceptibility of the material. The polarization with the $\cos 2\omega t$ dependence describes second harmonic generation.

This nonlinear optical process is not relevant to the generation and detection of THz radiation by nonlinear optical techniques and therefore is not discussed later.

Similarly, consider two optical fields oscillating at frequencies $E_1 = E_0 \cos \omega_1 t$ and $E_2 = E_0 \cos \omega_2 t$:

$$P_2^{nl} = \chi_2 E_1 E_2 = \chi_2 \frac{E_0^2}{2}[\cos(\omega_1 - \omega_2)t + \cos(\omega_1 + \omega_2)t]. \qquad (4.10)$$

In this case, the DC second-order nonlinear polarization P_2^{nl} consists of a term $P_2^{\omega_1 - \omega_2}$ proportional to the difference frequency $\omega_1 - \omega_2$ and a term $P_2^{\omega_1 + \omega_2}$ proportional to the sum frequency $\omega_1 + \omega_2$. The generation of the THz radiation by optical rectification relies on low-frequency generation and is described by the term $P_2^{\omega_1 - \omega_2}$. However, sum frequency generation $P_2^{\omega_1 + \omega_2}$ is not relevant to the generation of the THz radiation by nonlinear optical technique.

The same generation of THz pulses from optical rectification of fs laser pulses is based on difference frequency mixing of all frequencies within the bandwidth $\Delta\omega$ of an fs near-infrared laser pulse. Specifically, an fs pulse Δt is characterized by a large frequency bandwidth $\Delta\omega$. The bandwidth of the laser pulse is described by a Gauss function of width $1/4\Gamma$.

$$E(\omega) \propto \exp\left(-\frac{(\omega - \omega_0)^2}{4\Gamma}\right). \qquad (4.11)$$

In the time domain, the fs laser pulse is described by an optical field oscillating at frequency ω_0 with a time dependence described by a Gauss function:

$$E(t) = E_0 \exp(i\omega_0 t) \exp[-\Gamma t^2]. \qquad (4.12)$$

The bandwidth of the THz radiation pulse is determined by difference frequency generation by all frequencies within the bandwidth of the fs laser pulse. The time profile of the radiated THz pulse from optical rectification of the fs laser pulse is proportional to the second time derivative of the difference frequency term $P_2^{\omega_1-\omega_2}$. The time dependence of $P_2^{\omega_1-\omega_2}$ is determined by the Gaussian time profile of the optical laser pulse.

The most widely employed crystals for nonlinear optical generation and detection of THz radiation are ZnTe, GaP, and GaSe, ZnTe and GaP. They exhibit zincblende structure and $4\bar{3}m$ point group symmetry. GaSe is a hexagonal crystal with $6\bar{2}m$ point group symmetry. All three crystals are uniaxial. Uniaxial crystals have only one axis of rotational symmetry referred to as the c-axis or optical axis of the crystal. The efficacy of ZnTe, GaP, and GaSe for nonlinear optical THz generation and detection is rated by the linear electro-optic coefficient of the materials. The linear electrooptic coefficients for ZnTe, GaP, and GaSe are listed in Table 4.1.

Efficient generation and detection of THz radiation by optical rectification and the Pockels' effect require single crystals with high second-order nonlinearity or large electrooptic coefficients, proper crystal thickness and proper crystal orientation with respect to the linear polarization of the THz radiation. The surfaces of the crystals should be optically flat at the laser excitation wavelengths and of high crystalline quality (*e.g.* low levels of impurities, defects, intrinsic stresses, etc). The bandwidth of an electro-optical crystal for THz generation and detection is determined by the coherence lengths and optical phonon resonances in the material.

If the refractive indices of the new infrared laser excitation frequency and the THz radiation are identical, then the bandwidths of the THz radiation depend only on the pulse width of the incident fs near-infrared laser beam. Also, the strength of emission and sensitivity of THz detection

Table 4.1.   Electro-optical coefficients of some commonly known THz emitters and detectors.

Material	Structure	Electrooptic coefficient (pm/V)
ZnTe	Zincblende	$r_{41} = 3.9$
GaP	Zincblende	$r_{41} = 0.97$
GaSe	Hexagonal	$r_{22} = 14.4$
ZnSe	Zincblende	$r_{41} = 2.00$
$L_i TaO_3$	Trigonal	$r_{33} = 30.3$
DAST	Hexagonal	$r_{11} = 160$
Poled polymers	Hexagonal	$r_{11} = 160$

would increase similarly for all frequencies within the bandwidth with increasing crystal thickness.

However, the refractive indices for the near-infrared laser frequency and THz frequency are in general not the same. Therefore, electromagnetic waves at THz and near-infrared frequencies travel at slightly different speeds through the crystal.

The efficacy of nonlinear optical THz generation and detection decreases if the mismatch between the velocities becomes too large. The distance over which the slight velocity mismatch can be tolerated is called the coherence length. As a result, efficient THz generation and detection at a given frequency only occur for crystals that are equal in thickness or thinner than the coherence length for this frequency.

The THz emission strength of a crystal and sensitivity of an electrooptic THz detector are proportional to the thickness of the crystal. However, the bandwidth of the THz emitter crystal and bandwidth of the THz detector crystal also depend on the crystal thickness. In general, the THz emission strength and THz emission bandwidth for a crystal have a reciprocal relationship. The THz emission bandwidth increases when the crystal becomes thinner. The THz emission strength decreases when the crystal becomes thinner. The same rule applies to the sensitivity of an electrooptic THz detector and the detector bandwidth. Thinner crystals have a higher bandwidth but lower sensitivity than thicker crystals and vice versa.

For efficient generation of THz radiation by optical rectification of femtosecond laser pulses from zincblende structure crystals, it is important to select the proper orientation of the crystal with respect to the linear polarization of the laser beam.

In case of the longitudinal linear electrooptic effect, the electric field is applied perpendicular to the direction of propagation of the optical probe beam. Detection of THz radiation pulses is usually performed using the geometry of the transverse electrooptic effect. In general, for optical rectification to take place, several conditions must be met:

(1) The photo energy of the excitation beam is higher than the band gap of the material. In this case, phase matching is not important since the energy of the beam is absorbed within a short distance and the interaction range is much shorter than the coherence length;

(2) If the phase matching is satisfied, the generated THz field will continually increase within the crystal. Thus, strong THz generation

takes place. In this case, the phase matching is not satisfied, and the THz waves will survive only within the coherence length and the THz radiation will be weak.

(3) THz waves generated within a very short distance close to the surfaces of the crystal survive with velocity mismatch between the optical and THz waves. In this case, THz pulses will be comparatively weak. In practice, due to instability of the optical beam's wavelength, THz waves will be generated under all the above conditions.

The amplitude and phase of the sub-picosecond THz pulse are measured by recording the phase change of the fs near-infrared laser pulses through the electrooptic crystal. The electric field of the THz radiation induces a change of the index of refraction of the crystal via the linear electrooptic effect such that the material becomes birefringent. The phase retardation Γ between the ordinary and extraordinary ray after propagating through the birefringent crystal is proportional to the amplitude and phase of the THz electric field E, the crystal thickness l, the linear electrooptic coefficient r_{14}, the index of refraction of the crystal at the near-infrared laser frequency n_0, and the near-infrared wavelength λ:

$$\Gamma \propto \frac{1}{\lambda} \ln_0^3 r_{41} E_{THz}. \tag{4.13}$$

The phase retardation also depends on the orientation of the crystal and direction of polarization of the THz radiation pulse and the near-infrared pulse.

Table 4.2 shows the materials that are used for generation of THz optical rectification. Presented semiconductor materials are discussed further.

Table 4.2.   Materials used for generation of THz optical rectification.

Inorganic electrooptic crystals	Semiconductors	Organic electrooptic crystals
LiNbO$_3$	GaAs	4-N-methylstilbazolium togylate (DAST)
LiTaO$_3$	InP	N-benzyl-2-methyl-4-nitroaniline (BNA)
	CdTe	(-)2-(α-methylbenzyl-amino)
	InAs	-5-nitropyridine (MBANP)
	InSb	Electrooptic polymers
	GaP	
	ZnTe	
	ZnLdTe	
	GaSe	

4.2.1. *Semiconductor materials*

THz radiation generation in semiconductors was initially explained as the dipole radiation of a time-varying current resulting from photo-excited charge carriers in the depletion field close to the semiconductor surface. Although the ultrafast photocarrier transport model successfully explained the transverse magnetic polarization of the emitted radiation and the observed emission maxima for the Brewster angle incidence, the observed intensity modulation of the emitted THz radiation when the crystal was rotated about its surface, indicated that there were several mechanisms of the THz generation.

In 1992, Chuang *et al.* proposed a theoretical model based on optical rectification of femtosecond laser pulses at semiconductor surfaces, which successfully explained all of the earlier experimental observations. The investigation of the physics of optical rectification indicates that depending on the optical fluence, the THz radiation generation by optical rectification is either a second-order nonlinear optical process governed by the bulk second-order susceptibility tensor χ_2, or a third-order nonlinear optical process whereby a second-order nonlinear susceptibility results from the mixing of a static surface depletion field and the third-order nonlinear susceptibility tensor χ_3.

THz radiation generation from optical rectification of femtosecond laser pulses has been reported from $\langle 100 \rangle$, $\langle 110 \rangle$, and $\langle 111 \rangle$ oriented crystals with zincblende structure commonly displayed by most III–V and some II–VI compounds. A dramatic variation in the radiated THz signal along with polarity reversal is observed in $\langle 110 \rangle$ and $\langle 111 \rangle$ CdTe and GaAs crystals, as the incident photon energy is tuned near the band gap. This phenomenon can be explained as the dispersion of the linear susceptibility tensor near the electron resonance state.

THz generation by optical rectification is known to originate from narrow band gap semiconductors such as InAs and InSb at high optical fluence $(1$–$2\,\mathrm{mJ/cm^2})$ and off-normal incidence.

Experimental results of THz emission from $\langle 100 \rangle$ InSb crystal and $\langle 100 \rangle$, $\langle 110 \rangle$, and $\langle 111 \rangle$ InAs for both n-and p-type doping revealed a pronounced angular dependence of the emitted THz radiation on the crystal orientation. The results are in agreement with the theory that predicts that for high excitation fluence, THz emission from narrow band gap semiconductors results primarily from surface electric field-induced optical rectification.

Among the zincblende crystals, perhaps the most likely candidate for generation of THz waves by optical rectification is ZnTe. Nahata *et al.* first reported THz generation in ZnTe in an experiment that used a pair of ZnTe crystals for both generation and detection. In their work, a $\langle 110 \rangle$-ZnTe was pumped by 130 femtosecond pulses at 800 nm from a mode-locked Ti sapphire laser. The Fourier spectrum of the temporal waveform demonstrated a broad bandwidth with useful spectral information beyond 3 THz.

Apart from the zincblende crystals described earlier, GaSe is a promising semiconductor crystal that has been exploited recently for THz generation with an extremely large bandwidth of up to 41 THz (Huber *et al.*, 2000). It has a hexagonal structure and a direct band gap of 2.2 eV at 300 K. The crystal has a large nonlinear optical coefficient (54 pm/V), high damage threshold, suitable transparent range, and low absorption coefficient that makes it an attractive option for generation of broadband mid-infrared electromagnetic waves. For the broadband THz generation and detection using a sub-20 fs laser source, GaSe emitter-detector system performance is comparable to or better than that of thin ZnTe compounds. Using GaSe of appropriate thickness as emitter and detector, it is also possible to obtain a frequency-selective THz wave generation and detection system. The disadvantage of GaSe, although, is in the softness of the material that may exhibit fragility during operation.

4.2.2. *Inorganic electro-optical compounds*

In 1971, Yang and coworkers first demonstrated the generation of THz waves by optical rectification of picosecond optical pulses in organic electrooptic materials.

In their experiment, they observed THz radiation from a $LiNbO_3$ crystal illuminated by picosecond optical pulses from a mode-locked Nd: glass laser. In the late 1980s, Auston and co-workers reported THz radiation generation by optical rectification in $LiTaO_3$ with spectral bandwidth as high as 5 THz. A major drawback in their experiment was the difficulty in extracting the THz pulses from the crystal because of total internal reflection due to the small critical angle and high static dielectric constant in $LiTaO_3$. Later, Auston *et al.* described an approach that minimized the reflection loss and allowed the extraction of the THz waves. Zhang *et al.* later measured and calculated the optical rectification for $LiNbO_3$ under normal incident optical excitation. The maximum signal produced by $LiTaO_3$. was reported 185 times smaller than that emitted by a 4-N-methylstilbazolium tosylate (DAST) organic crystal.

4.2.3. *Organic electro-optical compounds*

Organic compounds have gained some popularity as THz emitters as it was reported that they generate stronger THz signals than commonly used semiconductors or inorganic electro-optical emitters due to their large second-order nonlinear electric susceptibility. Zhang *et al.* (1992) described THz optical rectification from an organic crystal, dimethyl amino DAST, which is a member of the stilbazolium-salt family. Electro-optical measurements at 820 nm have yielded reportedly a high electrooptic coefficient ($>$400 pm/V). In their work on THz emission by optical rectification of femtosecond laser pulses in DAST, Zhang *et al.* described a strong dependence of the radiated THz field amplitude on sample rotation about the surface normal, for both parallel and perpendicular orientations of the incident optical beam being pumped from the source. Zhang *et al.* (1992) also indicated that with a 180-mW optical beam (being pumped) focused into a 200-μm diameter spot, the best DAST sample provided detected THz electric field that was 185 times larger than that obtained from an $LiTaO_3$ crystal and 42 times larger than GaAs and InP crystals under the same experimental conditions. DAST has also been demonstrated to perform well at higher frequencies with an observable bandwidth up to 20 THz from a 100-μm DAST crystal showing a six-fold increase over that reported from a 30-μm ZnTe under similar conditions. The frequency spectrum of DAST shows a characteristic strong absorption line at 1.1 THz (from TO phonon resonance) along with some additional absorption lines between 3 and 5 THz. The latter (with the exception of the line at 5 THz) are significantly weaker than the absorption line at 1.1 THz, such that the THz amplitude at those frequencies is still substantially higher than the noise level.

Nahata *et al.* (1995) spoke out in favor of organic compounds in polymeric forms for THz generation instead of using single crystals. In particular, it is easier to process the former, they can be poled to introduce noncentersymmetry and have higher nonlinear coefficients than inorganic materials. Nahata *et al.* used a copolymer of 4-N-ethyl-N-(2-methacryloxyethyl) amine-4'-nitro-azobenzene (MA1) and methyl methacrylate (MMA) (commonly known as MA1:MMA) for generation of THz waves via optical rectification. The radiated field amplitude from a 16-μm thick sample (electrooptic coefficient $\sim$11 pm/V) was reported to be four times smaller than that observed from a 1-mm thick y-cut LiNbO$_3$ crystal. However, the coherence length of the polymer was reported to be 20 times larger. Nahata *et al.* proposed that the conversion efficacy relative to LiNbO$_3$ could be noticeably improved using available polymers with six times the nonlinear coefficient and millimeter-thick films poled at high field. This could be created by a dielectric stack of thin-poled films.

Although DAST has been reported performing better than ZnTe when used for THz generation by optical rectification, the THz electric fields generated by these crystals are more complicated in time and frequency domains. Phonon bands in these compounds also result in producing smaller range of frequencies than ZnTe.

We have thus reviewed THz optical rectification from a variety of materials including semiconductors and inorganic and organic crystals. The performance of these emitters in terms of bandwidths has been given in Table 4.3. ZnTe is the most popular choice because it can generate extremely short and high-quality THz pulses. Organic crystals have caused interest recently due to their higher electrooptic coefficient and enhanced capacity to generate stronger signals than commonly used ZnTe. However, as it was indicated before, the THz fields generated by organic crystals such as DAST and more recently BNA are more complicated in time and frequency domains. Low-frequency phonon bands also limit the bandwidth in these compounds compared with ZnTe. Organic compounds have also a large naturally occurring birefringence that complicates their application. This resulted in ZnTe remaining the preferable THz generator compound.

The choice of a suitable THz emitter also largely depends on the operating conditions. In order to achieve reasonable efficacy for nonlinear optical processes, long interaction length and appropriate phase matching conditions must be met. The phase matching condition is satisfied when the phase of THz waves travels at the group velocity of the optical pulse. In the case of ZnTe, the phase matching condition and the subsequent enhancement of coherence length are achieved at 800 nm, making it the

Table 4.3. Bandwidth performances of some common THz emitters (Wilke and Sengupta, 2007).

	Experimental bandwidth, THz	Lowest optical phonon resonance, THz at 300 K	Incident pulse wave-length, mm	Incident pulse width, fs	Detection scheme
GaAs	40	8.02	800	12	Electroop.
GaSe	41	7.1	780	10	—
ZnTe	17	5.4	800	12	—
GaP	3.5	10.96	1055	300	PC antenna
DAST	20	1.1	800	15	Electroop.
BNA	2.1	2.3	800	100	—

most suitable electrooptic compound for THz wave emission and detection using a Ti: sapphire laser system with a center wavelength of 800 nm.

Table 4.3 presents parameters of common THz emitters along with the corresponding detection schemes. Here, "Electroop" stands for electro-optical schemes and PC antenna for Photoconducting antenna scheme.

However, the large size of these lasers prevents them from being used in portable THz systems. In order to build a compact, integrated and sensitive THz imaging system, it could be necessary (as one of the possible options) to develop a THz system that would utilize the possibilities of the optical fiber.

The erbium (Er) and ytterbium (Yb)-doped fiber lasers operating at 1,550 nm and 1,000 nm, respectively, are especially attractive from the above-mentioned point of view. The large velocity mismatch and consequently a small coherent length, however, make ZnTe unsuitable for applications at these wavelengths. Although GaP and CdTe have been suggested as potentially compatible emitters for Yb-doped laser systems at 1,000 nm, GaAs is a more likely candidate for THz emission with Er-doped fiber lasers. Calculations show that the phase matching condition is satisfied at 1,050 nm and 1 THz in CdTe compounds resulting in an enhancement of coherence length whereas GaAs crystal phase matching occurs at 1,330 nm. This particular wavelength in GaAs is the longest among all binary semiconductors. Even at 1,550 nm, the coherence length still retains a large value, thus enabling GaAs to be used as a suitable emitter in this wavelength range. Another GaAs's merit is in its lowest phonon resonance lying at higher frequencies (10 THz) and consequently increased bandwidth capability. The coherence length of some organic crystals such as BNA is also quite large in the long wavelength regime; nevertheless, because of the difficulties mentioned before, organic crystals have not become widespread materials for THz emitters.

The THz emission from nonresonant optical rectification discussed earlier resulted from dipolar excitations for GaAs, ZnTe and $LiNbO_3$. THz emission may also result from nondipolar excitations. Emission via optical rectification of femtosecond laser pulses in $YBa_2Cu_3O_7$, single crystals of iron (Fe) and films consisting of nano-sized graphite crystallites was observed and was found to have resulted from quadrupole magnetic dipole nonlinearities.

For Fe, which crystallizes in a body-centered cubic lattice, optical rectification is forbidden by lattice symmetry. A nonvanishing second-order optical nonlinearity can, however, result from an electric quaropole magnetic dipole contribution, surface nonlinearity or sample magnetization. Each of the above contributions to the second-order nonlinear electric susceptibility has a characteristic dependence on sample azimuth and optical-pump pulse polarization for a given crystal orientation. The THz emission reported from Fe was found approximately three orders of magnitude weaker than from a 1-mm thick ⟨110⟩ ZnTe compound under the same optical excitation conditions.

THz emission from metals such as gold and silver has also received attention recently. Although second harmonic generation had been reported from metal surfaces as early as the 1960s, optical rectification of laser light had not been reported until recently. Kadlec *et al.* generated intense THz emission by optical rectification of *p*-polarized 810-nm laser pulses at a fluence of $6\,\mathrm{mJ/cm^2}$ in gold films. The THz amplitude was reported to scale quadratically with optical fluence for values up to $2\,\mathrm{mJ/cm^2}$. The *s*-polarized optical-pump beam caused THz emission, which was approximately three times weaker than for gold and showed linear dependence on incident pump fluence.

4.2.4. *Terahertz electro-optical detection*

Much research activity has been devoted to increasing sensitivity and bandwidth of electro-optical detection of THz-frequency radiation pulses. Other than the duration of the optical probe pulse, the bandwidth of free-space electro-optic sampling is only limited by the dielectric properties of the electrooptic compound or crystal. Electrooptic sampling of THz radiation is a powerful tool for detection of electromagnetic radiation pulses in and beyond the mid-infrared. A variety of electro-optic materials, including semiconductors and inorganic and organic electrooptic compounds, have been evaluated for THz detecting.

4.2.5. *Semiconductors and inorganic compounds for detectors*

As it was mentioned earlier, semiconductor and inorganic compounds remain to be popular choices for THz detectors. Below are some typical examples of implementations of devices with semiconductor and inorganic materials.

In 1995, Wu and Zhang demonstrated free-space electrooptic sampling of short THz radiation pulses (Wu and Zhang, 1995). In their work, a GaAs photoconductive emitter triggered by 150 fs optical pulses at 820 nm radiated electromagnetic waves of THz range. A 500-μm thick $LiTaO_3$ compound, with its *c*-axis parallel to the electric field polarization, was used as the electrooptic sampling element.

To improve detection efficacy, the THz radiation beam was focused onto the detector compound by a high-resistivity silicon lens. The temporal resolution of the detected THz transient was limited by the velocity mismatch between optical and THz frequencies. A signal-to-noise ratio of about 170:1 was obtained with a 0.3 second lock-in integration time constant.

Nahata *et al.* (1996) demonstrated a broadband time-domain THz spectroscopy system using $\langle 110 \rangle$ — oriented ZnTe compounds for THz generation by optical rectification and THz detection by electrooptic sampling.

An important condition for generation and detection of THz waves by nonlinear optical technique is the appropriate phase matching between the optical and THz pulses in the nonlinear optical crystal. For a medium with dispersion at optical frequencies, phase matching is achieved when the phase velocity of a THz wave is equal to the velocity of optical pulse envelope (or the optical group velocity). The coherence length l_c of ZnTe at optical wavelengths $\lambda = 780$ nm decreases.

The THz pulse and the optical probe pulse are incident onto the surface of the ZeTe compound parallel to the $\langle 110 \rangle$ plane. The optical axis of the crystal is in the $\langle 001 \rangle$ direction. The linear polarization of the THz pulse is perpendicular to the optical axis. The linear polarization of the incident optical probe pulse is parallel to the optical axis. The temporal waveform of the measured THz electric field and its Fourier amplitude and the Fourier plane are displayed in Figs. 4.4, 4.5, and 4.6 respectively. The peak of the THz pulse amplitude shows a three-fold rotational symmetry when ZnTe detector compound is rotated by 360° around an axis normal to the $\langle 110 \rangle$ plane. The result is inherent to all $\langle 110 \rangle$ and $\langle 111 \rangle$ zincblende crystals.

The broadband detection capability of *ZnTe* is obvious from the Fourier amplitude spectrum (Fig. 4.6), which extends well beyond 2 THz. The roll-off in sensitivity for higher frequencies from phase mismatch and the shorter coherence length are in agreement with the data given in Figs. 4.8, 4.9 and 4.11. A thinner crystal detects higher frequencies better than a thicker crystal. However, the detector sensitivity of a thinner crystal is reduced compared to a thicker crystal.

In addition to limitations imposed by phase mismatch, the absorption caused by a transverse optical phonon resonance at 5.4 THz in *ZnTe* is expected to attenuate the high-frequency components of the THz radiation. Although *ZnTe* delivers excellent performance in the 800-nm regime, its efficacy is inherently limited by large group velocity mismatch (GVM) of 1 ps/mm between optical and THz frequencies. An ideal material would be one with larger electrooptic coefficient and lower GVM. At 886 nm, GaAs

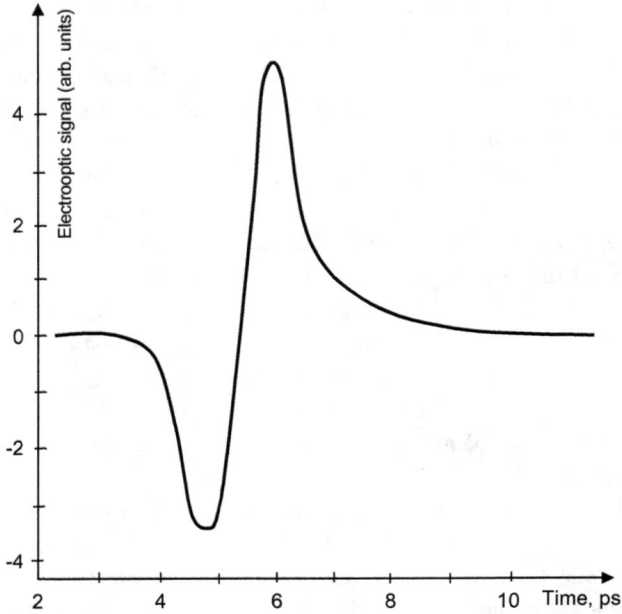

Fig. 4.5.   Time domain measurements of a THz pulse by electrooptic sampling with a ZnTe compound (Selig, 2000).

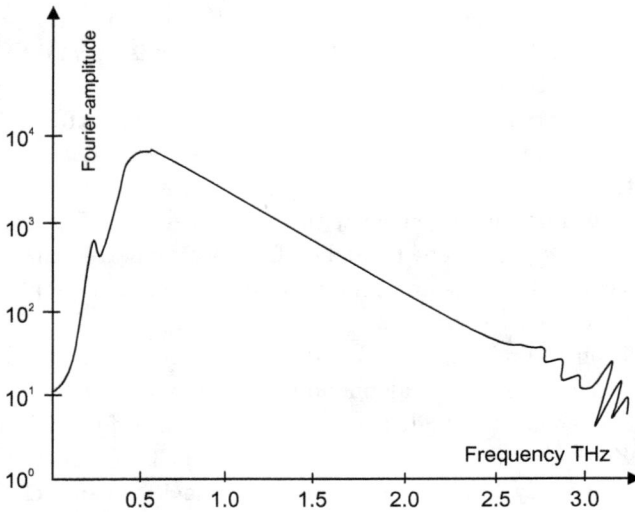

Fig. 4.6.   Fourier amplitude of the time-domain measurement displayed in Fig. 4.5. The bandwidth of the signal is about 2.75 THz. The dynamic range is 500:1 (Selig, 2000).

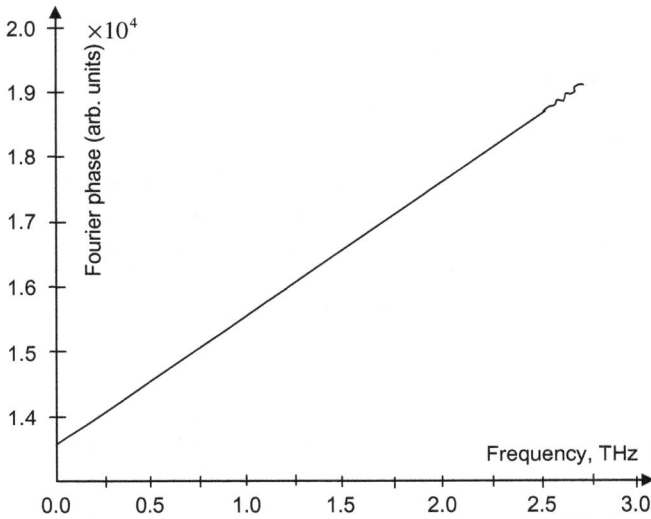

Fig. 4.7. Fourier phase of the time-domain measurement given in Fig. 4.4 (Selig, 2000). The linearity of the phase shows the linearity of the electrooptic response.

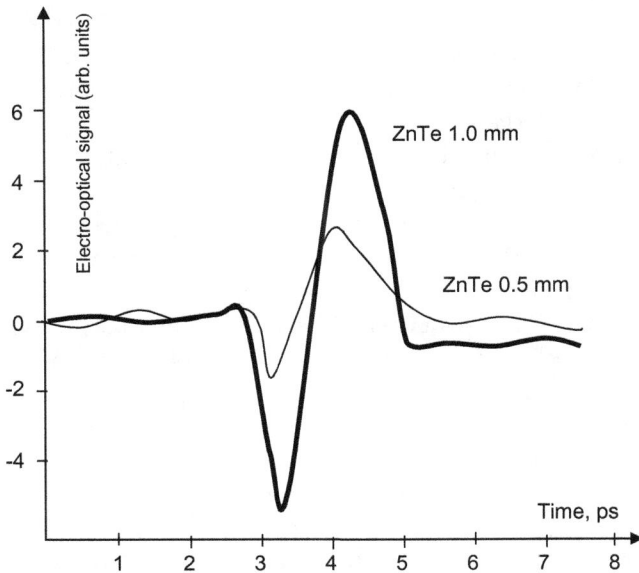

Fig. 4.8. Time-domain measurement of THz pulses by electro-optic sampling with *ZnTe* compounds of different thickness (Selig, 2000).

possesses a moderate electrooptic coefficient of 25 pm/V but a low GVM of 15 fs/mm. The electrooptic coefficient of LiTaO$_3$ is comparable to that of ZnTe but LiTaO$_3$ has a large GVM (>14 ps/mm), whereas for organic crystals (such as DAST) the electrooptic coefficient is very large but GVM is comparable to ZnTe's.

A substantial improvement in bandwidth detection and sensitivity may be obtained by using materials with phonon resonance occurring at higher THz frequencies. Among inorganic media with comparable nonlinear optical properties ZnSe (lowest TO phonon resonance at 6 THz), GaAs (TO phonon resonance at 8 THz), and GaP (TO phonon resonance at 11 THz) are good candidates.

Among zincblende compounds, GaP is a very likely alternative to ZnTe for electro-sampling due to its high-frequency phonon band at 11 THz, the highest among all zincblende compounds.

Figure 4.8 presents electrooptic sampling of a THz signal for ZnTe of different thickness. For both measurements, the optical probe beam and the THz pulse are both incident to the surface of the compounds parallel to the ⟨110⟩ plane. The surface of the compounds equals 10 × 10 mm. The linear polarization of the incident THz-pulse and probe pulse are perpendicular to each other. The polarization of the probe beam is parallel to the optical axis (⟨001⟩ direction). The ratio of the peak-to-peak amplitude of the THz pulses measured by the 1-mm compound and the 0.5-mm compound is 2.3 and scales approximately with the thickness of the two compounds (1 mm/0.5 mm = 2). The thicker ZnTe compound exhibits a larger signal than the thinner compound because the phase retardation $\Gamma \propto l$ is linearly proportional to the thickness l of the electrooptic compound. The electrooptic coefficient of GaP ($r_{41} \approx 0.97$ pm/V), however, is one fourth that of ZnTe.

One of the recently reported materials used for free-space electrooptic sampling of the THz waves is ZnSe in crystalline and polycrystalline form. Although the electrooptic coefficient of ZnSe ($r_{41} \approx 2$ pm/V) is only half of that of ZnTe, the TO phonon resonance frequency of 6 THz is higher than that of ZnTe (5.4 THz), thus promising a higher detection bandwidth potential for the former.

The GVM for ZnSe (0.96 pm/V) is comparable ZnTe. Using a 10-fs titanium–sapphire laser pulse to excite a 100-μm GaAs photoconducting antenna and 0.5 mm ⟨111⟩-oriented ZnSe single crystal as a detector, a spectral bandwidth of 3 THz was measured. In this case, the bandwidth detected by the ZnSe was limited only by the generation of the THz radiation. It was reported that for thicker (1 mm) polycrystalline electrooptic ZnSe, the random nature of crystallographic orientation within the interaction length distorted the phase of THz waveform. However, a reduction of the thickness to 0.15 mm minimized the distortion and extended the bandwidth up to 4 THz. Since polycrystalline semiconductors offer practical advantages for fabrication over their single crystal counterparts, the successful application of polycrystalline materials for free-space electrooptic sampling allows the possibility of utilizing nonlattice matched thin film integrated electrooptic detectors of THz radiation.

As it was mentioned earlier, the mismatch between the THz phase velocity and the group velocity of the optical probe pulse limits the detection bandwidth of zincblende electrooptic compounds such as ZnTe and GaP. However, this limitation can be eliminated by using a detection scheme, taking advantage of the type II phase matching in a naturally birefringent compound such as GaSe. The phase matching condition can be satisfied by angle tuning whereby the electro-optic crystal (z-cut GaSe) is tilted by an angle θ_{det} (the phase matching angle) about a horizontal axis perpendicular to the direction of time-delayed probe beam.

Table 4.4 gives a summary of the mentioned before characteristics of compounds. It is necessary to note that the values are of experimental nature and approximate.

Figure 4.9 illustrates the dependence of the bandwidth on the thickness of the compound. The thinner ZnTe compound has a slightly larger bandwidth than the thicker ZnTe. The calculated bandwidths of a 0.5 mm and a 1.0 mm compounds are 2.86 THz and 2.62 THz respectively.

In Fig. 4.10, the nonlinear crystal qualities are shown: indices of refraction as functions of the optical wavelength and THz frequency. Optical radiation at 760-μm wavelength propagates at the same speed as 2.6 THz

Table 4.4.   Properties of typical electro-optic sensor materials (Wilke and Sengupta, 2007).

Material	Crystal structure	Electrooptic coefficient (pm/V)	Group velocity mismatch (ps/mm)	Experimental detection bandwidth, THz
ZnTe	zincblende	$r_{41} = 3.90$	1.1	2.5
GaP	—	$r_{41} = 0.97$	—	7.0
ZnSe	—	$r_{41} = 2.00$	0.96	3.0
LiTaO$_3$	trigonal	$r_{41} = 30.3$	14.1	—
GaSe	hexagonal	$r_{41} = 14.4$	0.10	120
DAST	—	$r_{11} = 160$	1.22	6.7

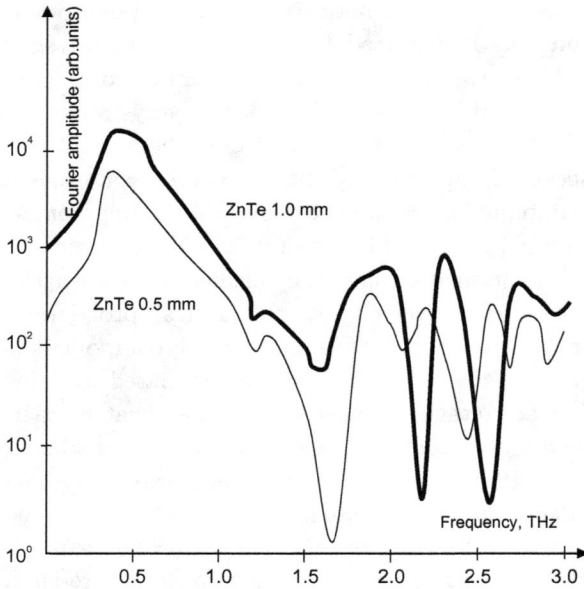

Fig. 4.9.   Fourier amplitudes of the time-domain measurements displayed in Fig. 4.6 (Selig, 2000).

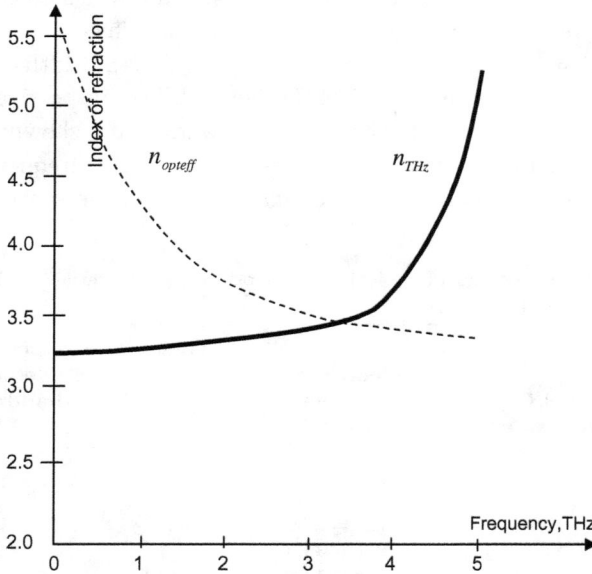

Fig. 4.10.   Index of refraction of ZnTe at optical wavelength (n_{opteff}) and THz frequencies (n_{THz})[9].

frequency radiation in *Zn Te*. The effective index of refraction at $\lambda = 780\,\mu$m is $n_{opteff} = 3.27$.

4.2.6. *Organic crystals*

Organic electro-optic crystals have attracted attention due to their high electrooptic coefficients that creates possibility of creating detectors with broad bandwidths. Han *et al.* (2000) demonstrated the application of the organic ionic salt compound DAST as a free space electro-optical detector of THz waves. The electro-optic coefficient for DAST is 160 pm/V at 820 nm and is almost two orders higher than that found in ZnTe (see Table 4.4). However, DAST exhibits two confirmed phonon absorption peaks at 1.1 THz and 3.05 THz.

Among other organic materials, poled polymers are known to produce good broadband electrooptic detection capability. These materials demonstrate large electrooptic coefficients, low dispersion between the THz and optical refractive indices, easiness of fabrication and easily modifiable chemical structures for better THz and optical properties because of their

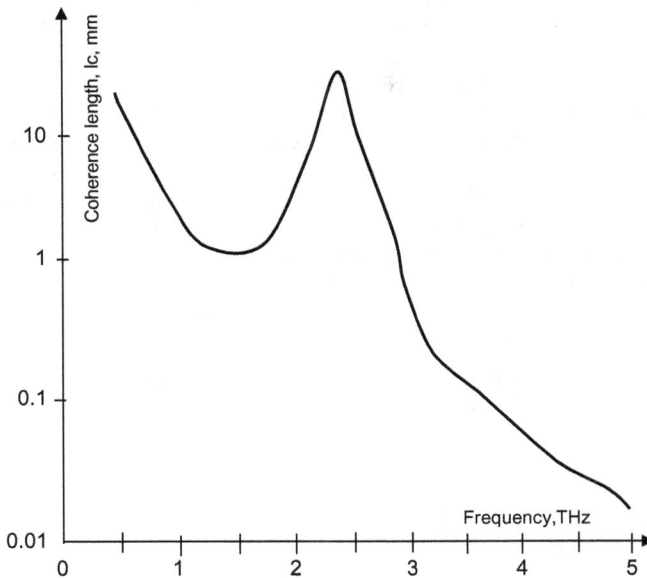

Fig. 4.11. Calculated coherence length l_c of ZnTe as a function of THz frequency for $\lambda = 780\,\mu$m (Selig, 2000).

organic qualities. Using poled polymers for free-space electro-optic detection, Nahata *et al.* recently measured an amplitude spectrum up to 33 THz.

As it was mentioned earlier, Fig. 4.11 shows a "roll-off" of coherence length for ZnTe as a function of frequency. The dependence is close to linear except for the range around 1.5 THz.

References

Han, PY, *et al.* (2000). Use of organic crystal DAST for terahertz beam applications. *Optics Letters*, 25, 675.

Huber, R, *et al.* (2000). Generation and field resolved detection of femtosecond electromagnetic pulses tunable to 41 THz. *Applied Physics Letters*, 76, 3191.

Nahata, A, *et al.* (1995). Generation of terahertz radiation from a poled polymer. *Applied Physics Letters*, 67, 1358.

Nahata, A, AS Weling and TF Heintz (1996). A wideband coherent terahertz spectroscopy system using optical rectification and electrooptic sampling. *Applied Physics Letters*, 69, 2321.

Selig, H (2000). Electro-optisches sampling von terahertz pulsen. *Hochschulschrift, Hamburg Universitaet, Fachbereich Physic Diplomarbeit*, 7.

Verghese, S, KA McIntosh, S Calana, WF Dinatale, EK Duerr and KA Molvar (1998). Generation and detection of coherent terahertz waves using two photomixers. *Applied Physics Letters*, 73(26).

Ward, J, E Schlecht and G Chattopadhyaya (2006). Capability of THz sources based on Schottky diode frequency multiplier chains. *IMS Publications (J. Ward)*, June.

Wilke, I and S Sengupta (2007). Nonlinear optical techniques for terahertz pulse generation and detection — optical rectification and electrooptical sampling. In *Terahertz Spectroscopy: Principles and Applications*, SL Dexheimer (ed.), Optical Science and Engineering, Vol. 131, p. 41. CRC Press.

Wu, Q and XC Zhang (1995). Free-space electrooptic sampling of terahertz beams. *Applied Physics Letters*, 67, 3523.

Zhang, XC, *et al.* (1992). Terahertz optical rectification from a nonlinear organic crystal. *Applied Physics Letters*, 61, 3080.

Chapter 5

Methods and Technology for THz Sources, Detectors and Processing Electronics

5.1. THz sources

There are many types of sources of terahertz (THz) radiation. THz are nothing special — a human body emits such waves and the matter around us does that too because we are all "blackbodies". However, the efficiency of a blackbody source is not very high. For example, the spectral power density of the sun is 20,000 times higher at visible frequencies than at 1 THz.

THz emitters can be realized using, e.g. microwave technology based on Gunn, Impatt, or resonant tunneling diodes. The fundamental frequencies of these devices, however, are typically not high enough for many THz applications and they need to be multiplied in special mixers. Although a microwave-based THz source can be fairly small (comparable to an average-size radio receiver), its cost can exceed several tens of thousand dollars.

The necessity to make THz sources more practical is one of the major tendencies at the moment. Below, there are several main types of THz sources available commercially and which can be reliably used for various applications.

The sources can be divided into the following:

(1) Continuous-wave (CW) and
(2) Pulsed sources.

One type of CW THz source is the optically-pumped THz laser (OPTL). (*Note*: OPTL may be used in a pulse-mode as well). OPTL lasers are in use around the world, primarily for astronomy, environmental monitoring as well as biological early warning systems and plasma diagnostics. Average power exceeds 100 mW in the range of 0.3–10 THz. They are commercially available and turnkey. These improved systems stem from several developments, including permanently scaled, single-mode, frequency-stabilized, folded-cavity, radio-frequency-excited waveguide CO_2 lasers; scaled FIR gas cells that eliminate gas transport issues; and exquisitely stable passive resonator structures.

The field of time-domain spectroscopy (TDS) typically relies on a broadband short-pulse THz source. The generation and detection scheme is sensitive to the sample material's effect on both the amplitude and the phase of the THz radiation. In this respect, the technique can provide more information than the conventional Fourier-transform spectroscopy that is only sensitive to the amplitude. Usually, the THz pulses are generated by an ultrashort pulsed laser and last only a few picoseconds. A single pulse can contain frequency components covering the whole THz range from approximately 0.05 to 4 THz. For detection, the electrical field of the THz pulse is sampled and digitized. Two common detection schemes are *photoconductive sampling* and *electro-optical sampling*.

Gyrotrons are high powered vacuum tubes that emit millimeter-wave beams by bunching electrons with cyclotron motion in a strong magnetic field. Output frequencies range from ~20 to 250 GHz, covering wavelength from microwaves to the edge of the THz band. Typical output power range spans from tens of kilowatts to 1–2 megawatts. Gyrotrons can be designed for pulsed or continuous (CW) operation.

The gyrotron is a type of free electron maser (microwave amplification by stimulated emission of radiation). It has high power at millimeter wavelengths because its dimensions can be much larger than the wavelength, unlike conventional vacuum tubes, and it is not dependent on material properties, as are conventional masers. The bunching depends on a relativistic effect called the Cyclotron Resonance Maser Instability. The electron speed in a gyrotron is slightly relativistic (comparable but not close to the speed of light). In this aspect, the gyrotron is different from the free electron laser that works on different principles and which electrons are highly relativistic.

As far as the military applications are concerned, gyrotrons are used in a weapon system intended for nonlethal crowd control that is called the Active Denial System, which delivers a sensation of intense heat to its target using a directional beam of energy.

Photomixers: Optical heterodyne conversion or photomixing is a technique that allows generating continuous wave (CW) radiation at THz frequencies using thin films of low-temperature-grown GaAs. Optimizing photomixers for maximum output power requires careful design of the epitaxial growth sequence, and detailed analyses of the radio-frequency (RF) circuits as well as of the optical feed. The necessity of control of the Low-Temperature-Grown (LTG) — GaAs epitaxy leads to materials with short photocarrier lifetime and robustness to thermal failure. Photomixers are compact, all-solid-state sources that are a pair of single-frequency tunable diode lasers used to generate a THz difference frequency by photoconductive mixing in LTG GaAs. The output frequency is tuned over the range of several THz by temperature or current detuning of the two diode lasers by a few nanometers in wavelength. As sources, photomixers have been used for local oscillators with cryogenic THz heterodyne detectors and for high-resolution gas spectroscopy when used in conjunction with liquid-helium-cooled bolometers. In simple terms, optical heterodyne conversion with a photoconductive switch (photomixing) is analogous to the operation of a transistor amplifier. In a transistor, a small signal applied to the gate (or base) modulates the conductance of a switch under a relatively large DC bias. The modulated output power is drawn from the source providing the DC bias. In photomixing, the photoconductance is modulated by the optical beating of the two laser diodes, and the output THz power is generated from current drawn from the battery that provides a DC bias between the photoconductor electrodes. Photomixing is fundamentally different from difference-frequency generation using a $\chi^{(2)}$ process in a material such as LiNbO$_3$. In this type of difference-frequency generation, the output THz power is generated from the optical photons, and only one THz photon can be created per pair of optical photons. The fact that two hot photons are required to create one cold photon results in an efficiency loss that makes photomixing more favorable only between 1 and 3 THz. At frequencies higher than several THz, however, the $\chi^{(2)}$-mixing process is more efficient than photomixing, since photomixers then suffer from parasitic impedance and the $\chi^{(2)}$-mixing process becomes more efficient.

Another source of CW is Backward Wave Oscillators (BWO). BWOs are electron tubes that can be used to generate tunable output at the long-wavelength end of the THz spectrum. These generators are based on high power, low-frequency BWOs combined with frequency multipliers. Output power of these devices decreases from 10–100 mW in 100–370 GHz range to 0.1–1 mW in 400–1100 GHz range and 0.001–005 mW at frequencies above 1100 GHz, but it remains well above detection limits of sensitive THz detectors (e.g. by Microtech Instruments Inc.).

The frequency multipliers are very compact devices based on Schottky diodes. The multipliers are attached to BWO's output via impedance matched waveguide adapters, enabling easy transition from one spectral range to another.

Direct multiplied (DM) sources are capable of producing milliwatts to microwatts of CW THz radiation power (decreasing with increasing of frequency) from 0.1 to 1 THz. They are turnkey and commercially available by a number of companies (e.g. Virginia Diodes, Inc.). DM sources take multimeter-wave sources and directly multiply their output up to THz frequencies. DM sources with frequencies up to a little more than 1 THz and approximately $1\,\mu W$ of output have been used as local oscillators for heterodyne receivers in select applications. However, they can produce substantially more output power at lower frequencies, and they are often well suited to applications that require less than 500 GHz.

At the present moment, there are THz sources that are still at the experimental stage of development. One of them is Quantum-cascade lasers operating at wavelengths in the 4.4 THz regime. These lasers are made from 1,500 alternating layers of GaAs and AlGaAs and have produced about 2 mW of peak power (20 nW average power), and advances in output power and operating wavelength continue at a rapid pace. These lasers currently work best at only a few Kelvins, but in the future, they could become an improvement source of commercial THz systems.

In addition to the electronic means provided previously, there are a number of other devices used for generation, detection and processing of THz radiation (see Ch. 4 for details on the physics of operation) such as bolometers and heat detectors cooled to liquid-helium temperatures. However, since bolometers can only measure the total energy of a THz pulse, rather than its electrical field over time, it is not suitable for use in THz-TDS.

The materials used for generation by optical rectification can also be used by using the Pockels effect, where certain crystalline materials become birefringent in the presence of an electric field of a THz pulse leads to a change in the optical polarization of the detection pulse, proportional to the electric field strength. With the help of polarizers and photodiodes, this polarization change is measured.

Figure 5.1 presents a diagram that allows visualizing peak power and frequency ranges of different sources of THz waves.

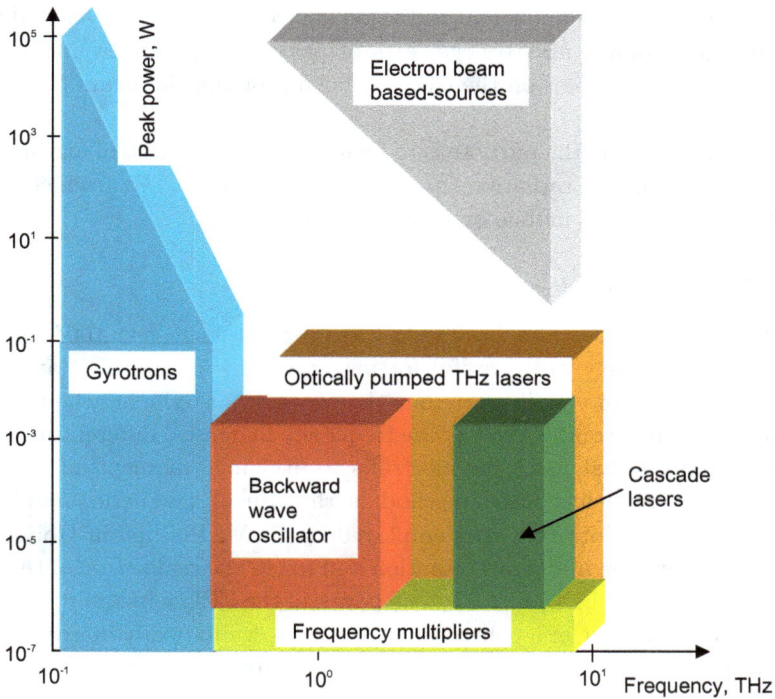

Fig. 5.1.   THz sources: Frequency and power ranges diagram.

5.1.1. *Pulsed THz sources*

5.1.1.1. *Example: Electron beam THz source*

Electron beam-based sources show promise as efficient and powerful THz devices (Biedron *et al.*, 2007). A number of laboratories have demonstrated THz radiation using electron beams and various radiation emission schemes. Indeed, THz radiation has long been used as a diagnostic of beam bunch lengths. Electron beam sources can either use short (∼100s of femtoseconds) bunches to directly produce THz (by either bending the electron beam or by striking a transition radiation target), or can produce pulses of arbitrary lengths by a collective radiation scheme, such as a free electron laser or the Smith–Purcell effect. At the present moment, electron-beam sources have already delivered several orders of magnitude, more peak and average power than other coherent THz sources. Also, the promise to provide still higher power in compact electrically efficient packages makes the electron-beam sources more attractive for military and special applications.

Recently, compact coherent radiation devices have been developed that are able to produce up to megawatts of peak THz power by inducing a ballistic bunching effect on the electron beam, forcing the beam to radiate coherently.

In this example, the purpose is to focus on the generation of high-power, continuously tunable radiation through the mm-wave to THz from compact electron sources as a mobile or stationary application.

5.1.1.2. *Background of electron beam sources*

Traveling-wave tubes and backward-wave oscillators in the 100 GHz and 1.2 THz respectively prove useful for low average ($\sim$1–100 mW) and relatively low-peak powers. To overcome the necessity of reducing the physical size of the source components as the frequency increases, different methods were developed that force the electrons to exchange momentum for THz emission. A frequently used method is the use of magnetic undulators proposed by H. Motz that was employed by R. M. Phillips in Ubitron in 1960 for generating mm-wave radiation and led to the realization of the first free-electron laser (FEL) in 1977. The power of the FEL is its ability to force the electrons to emit in a coherent fashion thus dramatically increasing the overall power of the emission. Other free electron coherent emission FEL are the Cherenkov FEL, based on the interaction with a dielectric loaded waveguide, and the metal grating based on the Smith–Purcell effect.

In his work on undulator radiation, H. Motz pointed out that in a uniform electron beam the contributions of individual electrons to the radiation field are random in phase, and therefore, the square of the total field equals the quadrature sum of the individual fields. However, if the electrons were bunched within a distance comparable to the wavelength of the radiation, their fields would add up in phase, resulting in the emission of coherent radiation in the THz region utilizing bunched electron beams, such as those produced by radio frequency (RF) accelerators.

More recently, the coherent spontaneous emission from an RF-modulated electron beam at wavelengths comparable to the electron bunch length has been the object of renewed interest because of its relevance to the generation of short pulses of coherent THz radiation.

5.1.1.3. *Source of radiation*

The principle of operation of the source is based on the coherent spontaneous emission from short bunches of relativistic electrons. The FEL source utilizes a 2.5 MeV RF linear particle accelerator (linac) to generate the electron beam, which is injected into a linearly polarized magnetic undulator composed of 16 periods, each 2.5 cm long with a peak magnetic field of 6000 Gauss. A second RF structure, called the phase matching device (PMD) is placed between the lunac and the undulator and is controlled in phase and amplitude to correlate the electron distribution in energy as a function of time in the bunch. Thus, the contributions to the total radiated field by individual electrons in the bunch are added in phase, loading to a manifold enhancement of the coherent emission.

5.1.1.4. *Undulator emission*

As it was mentioned before, the idea of undulator emission that allows producing high energy concentration is realized by periodic amplification of the beam power. Periodic enhancement centered at a specific frequency is achieved by using multiple, evenly spread, equal strength, alternating field magnets — undulators or wiggler magnets. The resonant wavelength for the output radiation is given by

$$\lambda_{\mathrm{rad}} = \frac{\lambda_0}{2\gamma^2}\left(1 + \frac{K^2}{2}\right), \tag{5.1}$$

where γ is the normalized electron beam total energy, λ_0 is the spatial period of the undulator, normalized undulator parameter is

$$K = 0.934\lambda_0[cm]B_{\mathrm{max}}[T]. \tag{5.2}$$

5.1.1.5. *Results*

The time-average power for the source is 64 mW. The introduction of cathode laser emission gating increased the beam currents by a factor of 100, leading to peak THz powers around a MW not compensating for losses through the system.

A maximum emitted power of about 1.5 kW in a 5-μs pulse duration has been measured at the peak of the phase-tuning curve when the RF field in the PMD (E_{pmd}) was set to about 0.5 the field in the lunac (E_{lunac}). The central wavelength of the emission in these operating conditions is 760 μm (0.4 THz).

5.1.2. *Semiconductor sources of THz radiation*

Many THz device applications require small-size, small-weight, low-consumption THz sources. However, semiconductor sources that seem like a good choice are difficult to implement due to the high-frequency implications of the THz range. A significant breakthrough in this direction took place in 2002 with the demonstration of a quantum cascade laser (QCL) operating at ∼4.4 THz. QCLs rely on transitions between quantized conduction band states of a suitably designed semiconductor multi-quantum-well structure. They are in-plane emitters, with the electric field vector perpendicular to the plane of the layers. The surface emission should be made possible. The advantage of transverse magnetic polarization in this case is realized by exploiting surface plasmons for waveguiding (plasmons can be described in the classical picture as an oscillation of free-electron density with respect to the fixed positive ions in a metal).

One of the possible implementation of the above type of a THz source is the use of photonic crystal structures as a basis for a semiconductor laser in the THz range. Semiconductor lasers based on two-dimensional (2-D) photonic crystals generally rely on an optically pumped central area surrounded by unpumped and, therefore, absorbing regions. This almost ideal configuration is useless when photonic-crystal lasers are electronically pumped (which is done in attempt to avoid using an external laser source). In this case, in order to avoid lateral spreading of the electrical current, the device active area must undergo special semiconductor processing. This creates an abrupt change in the complex dielectric constant at the device boundaries, especially for the lasers operating in the far-infrared, where the large emission wavelengths impose device thicknesses of several micrometers. Such boundary conditions can dramatically influence the operation of electrically pumped photonic-crystal lasers.

In the mid-infrared, photonic-crystal-based quantum cascade structures have been demonstrated for surface emission and normal-incidence detection. The application of photonic-crystal technology to active devices is advantageous since it allows the achievement of simultaneous spectral and spatial mode implementations. In the above-mentioned mid-infrared structures, the effective index contrast was achieved by deep etching of the semiconductor material. This approach has been transferred to the THz range to realize devices based on lattices of deeply etched pillars (Dunbar *et al.*, 2005). Besides the high degree of technological complexity, these devices did not provide surface emission and the possibility of building the photonic structure on the top metal only.

In the example below, the physical and technological principles that help successfully implement a photonic-crystal semiconductor laser as a compact source of THz radiation are illustrated. In general, photonic crystals are composed of periodic dielectric or metallo-dielectric nanostructures that affect the propagation of electromagnetic waves in the same way as the periodic potential in a semiconductor crystal affects the electron motion by allowed and forbidden electronic energy bands. Essentially, photonic crystals contain regularly repeating internal regions of high and low dielectric constants.

5.1.2.1. *Example: Electrically pumped photonic-crystal THz laser*

In the described example, the authors introduced a new approach to the implementation of a resonator for THz semiconductor lasers (Chassagnaeux *et al.*, 2009). It is based on a 2-D photonic-crystal structure, lithographically transferred onto the top metallization of a metal–metal THz quantum laser. In addition, it is shown that the operation of the photonic crystal is critically dependent on the boundary conditions. It seems to be a key element for the development of any electrically pumped photonic-crystal device. The described micro-resonator exhibits lithographically tunable single-mode emission pattern.

The quantum laser used in this work was grown by molecular beam epitaxy in the GaAs/AlGaAs material system. It consists of a 12-μm-thick, bound-to-continuum active region with emission at $\sim$2.7 THz. The active core is sandwiched between 700 nm, 2×10^{18} cm^{-3} doped lasers, forming the lower and upper contacts. The quantum cascade wafer was thermo-compressively bonded to $n^+ -$ GaAs wafer. After selective etching of the GaAs substrate, the 700-nm-thick top doped layer was thinned to 200 nm to reduce free-carrier absorption. Hexagonal mesas were etched down to the bottom metal to avoid lateral current dispersion, and the photonic-crystal pattern was implemented in the top metallization, which acts as a contact and, simultaneously, provides the necessary optical feedback. In Fig. 5.2(a), a schematic cross-section of the device is given.

The active region of the laser is located between two *Ti/Au* metal layers. The top metal is patterned with the desired photonic-crystal design. The top $n^+ -$ GaAs contact layer, approximately 200 nm thick, is removed in the photonic-crystal holes in order to reduce the losses. At the edge of the device $n^+ -$ GaAs layer can be removed or left in place, if mirror or absorbing boundary conditions, respectively, need to be implemented.

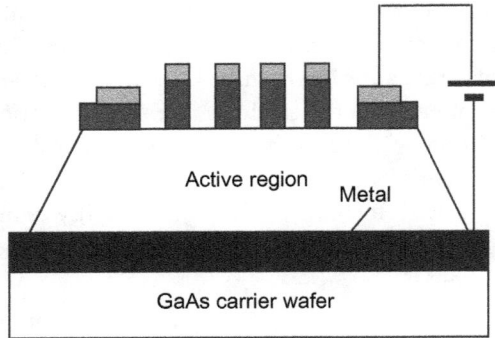

Fig. 5.2.   (a) Schematic cross-section of the device (Dunbar *et al.*, 2005).

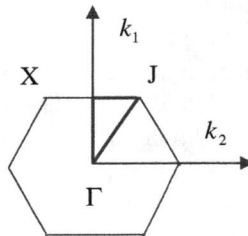

Fig. 5.2.   (b) (Part I) Transverse-magnetic photonic-crystal band structure (around point Γ) (Dunbar *et al.*, 2005).

Figure 5.2(b) presents a transverse-magnetic photonic-crystal band structure of the trigonal lattice used for the described example where $r/a = 0.22$. The calculations for the example were done in 3-D and did not need an effective index for the holes. The distribution of the electromagnetic field was calculated with a finite-element solver for a single lattice unit cell using Bloch-periodic boundary conditions. k_1, k_2 indicate axes in reciprocal space.

In the presented example, the authors implemented a band-edge-mode laser. In general, 2-D photonic-crystal slab waveguide lasers can be divided into two groups: defect mode lasers and band-edge mode lasers. The former operate at frequencies inside the bandgap by intentionally introducing a defect that supports localized modes. Band-edge mode lasers instead operate in regions of energy-momentum space that have a high photonic density of states and a corresponding small group velocity. The latter device architecture was implemented in order to take advantage of the connected nature of the lattice that significantly simplifies the processing, and the

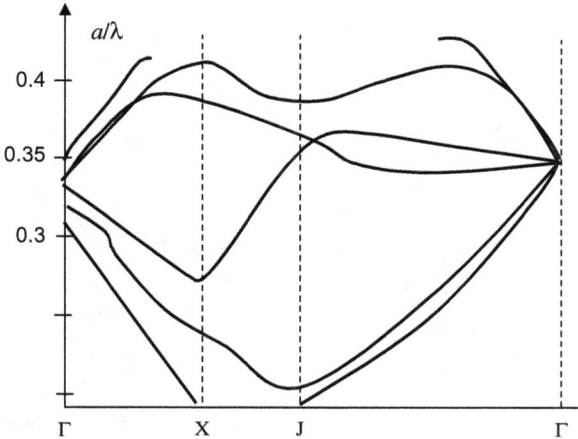

Fig. 5.2. (b) (Part II) Transverse-magnetic photonic-crystal band structure (around point Γ) (Dunbar *et al.*, 2005).

spatial delocalization of band-edge modes that allows an improved power extraction.

The boundary conditions represent a crucial point that needs to be addressed for a correct design. This issue is often overlooked for optically-pumped devices, as the photonic-crystal is usually much larger than the laser excitation spot. Absorbing boundary conditions are, therefore, naturally implemented due to the photonic crystal regions that are not optically pumped. The case of a current injection device is substantially different, since, in general, the photonic crystal must have a finite size. For metal–metal waveguides, in particular, it can be shown that the facet reflectivities are extremely high, as the waveguide thickness is much smaller than the lasing wavelength.

In Fig. 5.3(a), results for devices with absorbing and with mirror (the curves almost coincide and not shown separately for simplicity) boundary conditions are shown. A He-cooled Si-bolometer was used for detection. In Fig. 5.3(b), light current density and voltage current density curves are given as a function of **a** (in μm) for devices with $r/a = 0.22$. Devices with absorbing boundary conditions (A devices) are reported in (b) (dashed line in (b)), since its output power is 50 times reduced with respect to the others. The M (i.e. corresponding to $\underline{M}$irror boundary conditions) devices exhibit identical J_{th}, confirming that they are unaffected by the photonic structure patterned on the top metal, contrary to A (i.e. corresponding to $\underline{A}$bsorbing

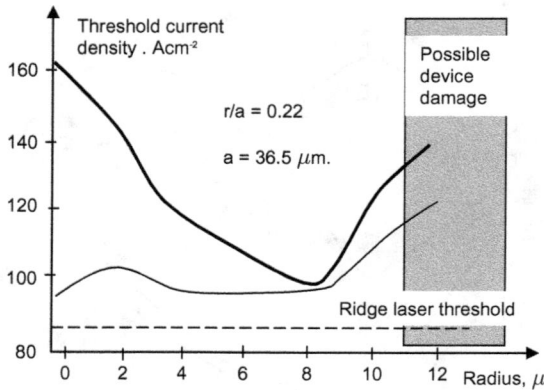

Fig. 5.3.   (a) Threshold current density ($T = 10\,\text{K}$) as a function of the photonic-crystal hole radius (r) and constant photonic spacing.

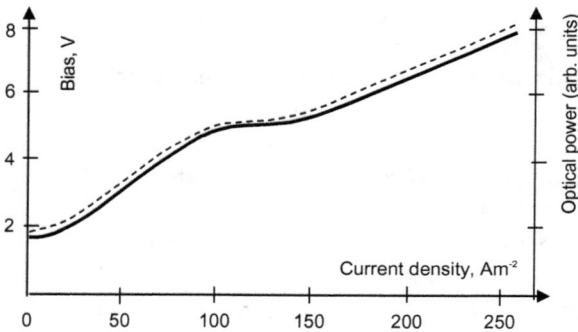

Fig. 5.3.   (b) Light current and voltage current density curves.

boundary conditions) devices. The threshold densities are calculated as follows: when the n^+ layer is present, the whole disk surface is employed, and when the n^+ layer is missing, the metal contact surface only is employed, since the top n^+ layer is responsible for the spreading of the current.

5.1.3. *Continuous wave sources*

Many sources are available for generating coherent continuous-wave THz radiation, including backward-wave oscillators, molecular lasers, Schottky diode upconverters, and parametric downconverters. If we restrict the discussion to solid-state devices, the most widely used sources are varactor multipliers or fundamental sources such as negative-resistance diodes and, more recently, high-electron mobility transistors (HEMTs). In general, electronic technologies that use electron transport for THz generation are limited by parasitic resistance and capacitance as frequency exceeds 1 THz. A developing technology is the THz QCL — only electroluminescence has been the major output of the device. However, this technology represents a fundamentally different approach to THz generation than the electronic technologies. Quantum effect devices such as Bloch oscillators and QCLs are generally limited to operation at cryogenic temperatures since thermal energy at 300 K otherwise smears out the quantum levels.

Photomixers may be thought of as combining some of the attributes of electronic and optical technologies. Although their output power is modest ($<10 \, \mu$W), this power is suitable or almost suitable for such applications as local oscillators and spectrometers. Typical optical-to-electrical conversion efficiency for photomixers is below 10^{-5} for a single device and output power falls from $\sim 2 \, \mu$W at 1 THz to below $0.1 \, \mu$W at 3 THz (Brown, 2003). Two of the photomixer's most attractive features are: its wide tuning range (25–3000 GHz), and that it operates at room temperature. Photomixers are promising continuous-wave THz sources. THz photomixing is an optical heterodyning scheme, in which optical modes generated by two single-mode lasers, or a dual-mode laser with their frequency difference fall in the THz range, mix in a photoabsorbing medium (heterodyne detection utilizes a nonlinear device called a "mixer"). This mixing process in the presence of an appropriate bias generates a THz signal with a frequency equal to the frequency difference of the two optical modes. The generated THz signal can be coupled to an integrated on-chip waveguiding structure or to a radiating antenna. The frequency of the THz signal can be tuned by changing the wavelengths of the lasers. An antenna coupled photomixer source is made of an ultra-fast photoconductor material with a pattern of closely spaced electrodes on it connected to an antenna structure. A DC bias is applied to the photoconductor through the electrodes. Two CW laser beams combined either inside an optical fiber or properly overlapped in space are focused on the biased region to generate a modulated carrier density, which generates a THz current in the presence of a bias voltage (McIntosh *et al.*, 1996).

5.1.3.1. *Example: Photomixer as CW, design features*

The basic design tradeoffs for lumped-element photomixers are limited by: optical spot size, thermal limits from optical and Ohmic heating, capacitance, and the photoconductor's RF resistance (Sakai, 2005).

A photomixer comprises a thin photoconductivity film of low-temperature-grown GaAs, submicron electrodes, and an antenna (also used to provide a DC bias). In the photoconducting film, the incident light modulates the conductivity through the generation of photocarriers. The photomixing process occurs during illumination of an electrode region with two single-mode lasers with average powers P_1 and P_2 and frequencies ν_1 and ν_2 respectively. The instantaneous optical power incident on the photomixer contains mixing terms (beats) between the two optical frequencies:

$$P_i = P_1 + P_2 + 2\sqrt{P_1 P_2} + 2\sqrt{P_1 P_2}(\cos 2\pi(\nu_1 - \nu_2)t + \cos 2\pi(\nu_1 + \nu_2)t),$$
$$(5.3)$$

where $f = \nu_1 - \nu_2$ is the difference frequency which can be easily adjusted by changing the bias current or operating temperature of the diode lasers. The $\nu_1 + \nu_2$ term occurring at the sum frequency does not couple efficiently through the THz metal circuit and may be neglected.

The DC photocurrent generated by the optical power is:

$$I_0 = \eta_e \frac{e}{h\nu} P_0,$$
$$(5.4)$$

where $e/h\nu$ is a constant of 0.69 for an optical wavelength near $860\,\text{nm}$ (corresponds to the band gap of GaAs) and η_e is the external quantum efficiency. Using the above two equations, the amount of THz power transmitted out of the antenna is then:

$$P_\omega = \frac{0.5 I_0^2}{1 + \omega^2 \tau_e^2} \eta_{\text{ant}} \text{Re}(z_{\text{ant}})$$
$$(5.5)$$

where $\omega = 2\pi f$, τ_e is the photocarrier lifetime, η_{ant} is an antenna efficiency that accounts for rf absorption, and $\text{Re}(z_{\text{ant}})$ is the real part of the antenna impedance and includes all parasitic reactances associated with the electrodes. By reducing τ_e, the optical power and DC bias can be increased thereby increasing the THz output power while staying below the maximum sustainable value of I_0. Figure 5.4(b) shows the maximum THz power available given the constraint on total I_0. Note that if you want to increase the diode laser power, a gain of 40 dB is achievable by reducing τ_e by a factor of 0.01. The optimum lifetime τ_e for a given operating frequency f is approximately the value of τ_e that satisfies the equation $2\pi f \tau_e \approx 1$. This represents a near optimal tradeoff between reduced power dissipation and efficient generation of THz photocurrent. Figure 5.4(a)

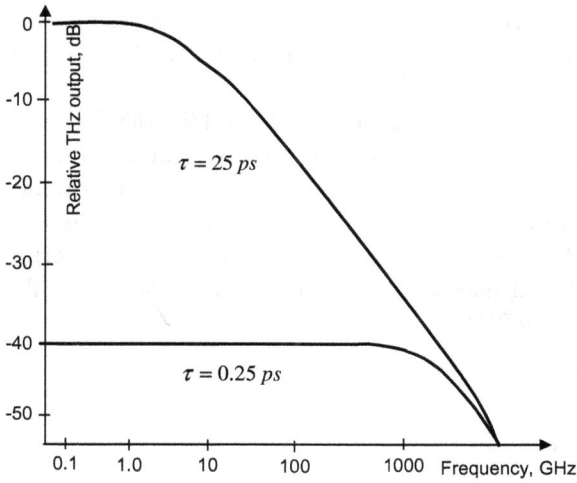

Fig. 5.4. (a) Output power dependence on different values of carrier lifetime.

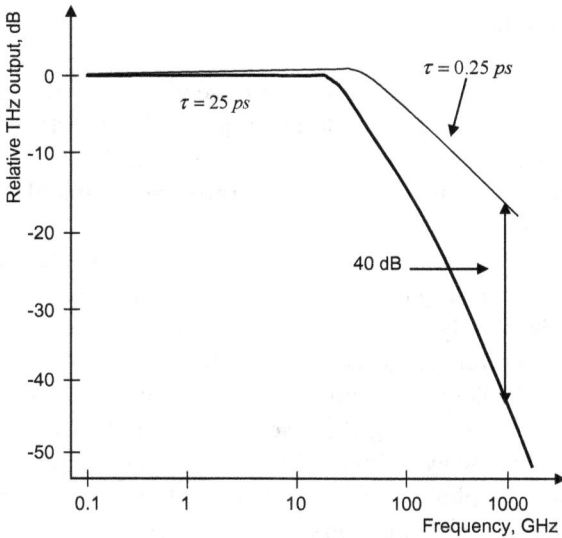

Fig. 5.4. (b) Two photomixers with different τ_e operating at the same DC photocurrent but with different laser illumination powers. These photomixers have similar amounts of Ohmic heating due to DC photocurrent but have very different THz output powers.

provides a comparison of the two output power values for photomixers with photoconductors with different values of carrier lifetime τ_e illuminated by the same laser power.

The low values of output power pose difficulties for a broad spectrum of photomixer applications. Nevertheless, photomixers have already been useful for local oscillators and gas spectroscopy. It is important to take into consideration the photocarrier lifetime and an antenna design in order to fabricate high-performance photomixers. Newer structures that use a distributed design may ultimately outperform conventional structures for frequencies above 2 THz.

5.2. THz detectors

THz detection schemes are largely classified as either coherent or incoherent techniques. The fundamental difference is that coherent detection measures both the amplitude and phase of the field, whereas incoherent detection measures the intensity. The incoherent intensity measurements are rather secondary in the light of THz identification and will not be considered in detail.

Coherent detection techniques are closely associated with generation techniques in that they share underlying mechanisms and key components. In particular, optical techniques utilize the same light sources for both generation and detection. Figure 5.5 illustrates the commonly used coherent detection schemes.

Free-space electro-optic (EO) sampling measures the actual electric field of broadband THz pulses in the time domain by utilizing the Pockels effect, which is closely related to optical rectification. In this case, a THz field induces birefringence in a nonlinear optical crystal, which is proportional to the field amplitude. The entire waveform is determined by a weak optical probe measuring the field-induced birefringence as a function of the relative time delay between the THz and optical pulses.

Sensing with a photoconducting (PC) antenna also measures broad-band THz pulses in the time domain. In the absence of a bias field, a THz field induces a current in the photoconductive gap when an optical probe pulse injects photocarriers. The induced photocurrent is proportional to the THz field amplitude. The THz pulse shape is defined in the time domain by measuring photocurrent while varying the time delay between the THz pulse and the optical probe.

A combined setup of broadband THz generation and detection measures changes in both the amplitude and phase of the THz pulses induced by a sample, which provides enough information to simultaneously determine the absorption and dispersion of the sample. This technique is called *THz time-domain spectroscopy* (THz-TDS) and was discussed in detail in Ch. 4.

Photomixing measures CW THz radiation by exploiting photoconductive switching. In this case, the photocurrent shows sinusoidal dependence on the relative phase between the optical beat and THz radiation. Using coherent laser beams to generate and detect THz signals, one can control the phase distribution of the generated THz signals in a photomixer antenna to increase the efficiency of the antenna and control the radiation beam.

Among different ways to detect THz signals, several types of devices are widespread and currently available from manufacturing companies (e.g. Microtech Instruments, Inc.).

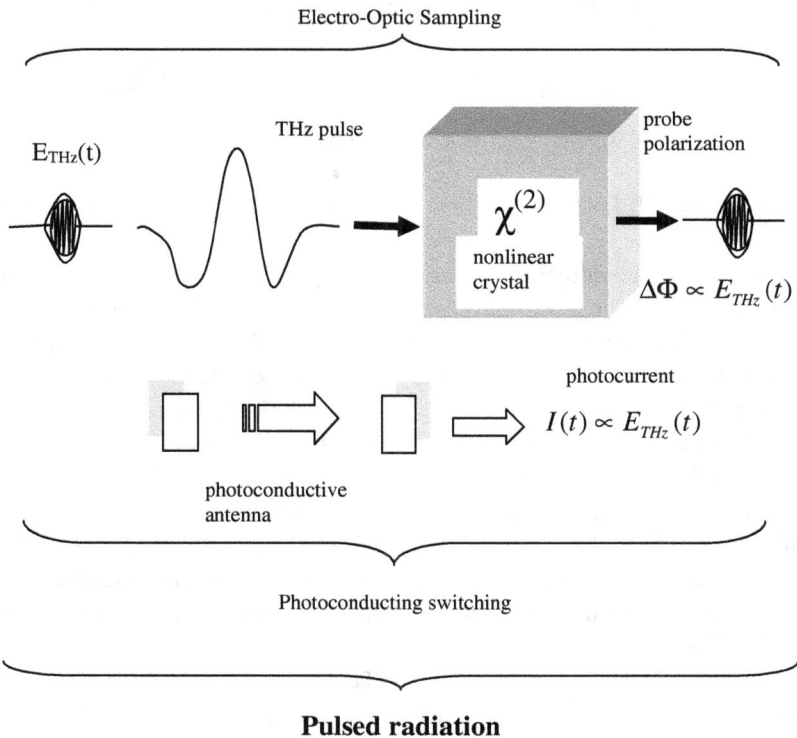

Fig. 5.5. (Part I) Coherent detection of THz radiation (pulsed radiation and CW).

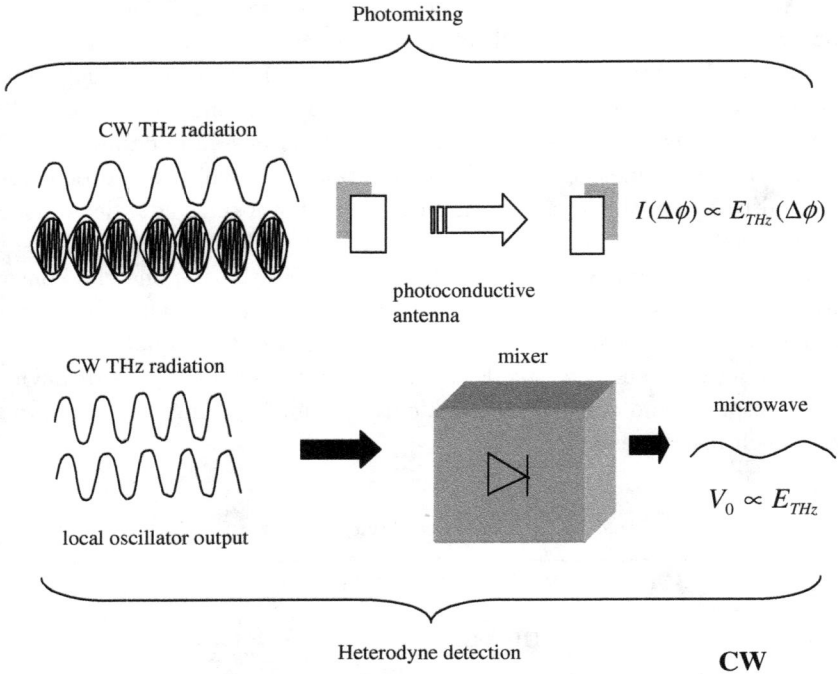

Fig. 5.5.   (Part II) Coherent detection of THz radiation (pulsed radiation and CW).

Opto-acoustic detector (Golay Cell) operates in the range of 0.02–20 THz, at ambient temperatures, and has a broad spectral response. It has a polyethylene input window that provides high transparency at frequencies up to 20 THz. A small fragile gas chamber includes a thin, partially absorbing film, and what is called an "optical microphone section". When the thin film in the gas cell absorbs IR or THz radiation, the gas is heated, and then it expands and distorts the mirrored back wall of the cell. This distortion (or movement) is monitored and measured by the combination of an LED, optics, grating and photodiode. The output of the photodiode is proportional to the displacement of the mirrored wall of the gas cell. Its output is calibrated against a source of known power output in Volts/Watt.

Pyroelectric Detector based on $LiTaO_3$ crystal is designed for registration of modulated electromagnetic radiation in the millimeter and submillimeter wavelength range (0.02–3 THz). At square wave modulation of radiation, the saw-tooth signal voltage from the detector is proportional to the intensity of radiation. Typical responsivity is about $1000\,\text{V/W}$ at dynamic range of $1\,\mu\text{W}$–$10\,\text{mW}$. The power supply current is on the order of 0.1–$0.2\,\text{mA}$, which makes it suitable for portable applications.

Currently, there are other detecting devices being developed. One is GaN/AlGaN transistors with high electron mobility in the 0.2–$2.5\,\text{THz}$ range (much higher than the cutoff frequency of the transistors). For the lowest temperatures, a resonant response was observed. Nonresonant detection was reported at temperatures above $100\,\text{K}$.

A number of steps have been taken to minimize the size and power consumption of the detectors: one was to develop a high-sensitivity THz detector on a chip. The detector itself consists of thin semiconductor films with a highly mobile layer of electrons that efficiently absorb THz radiation.

5.2.1. *Quantum superlattice as a THz detector*

One possible solution to create a sensitive THz detector is to exploit the extremely strong nonlinear dispersion relation for the electrons in the limit of a single miniband of a semiconductor superlattice. This idea was first proposed in 1997 by A.A. Ignatov *et al.*, when he and his co-authors calculated a DC current induced by the weak incident THz radiation (Fig. 5.6) and estimated the responsivity of such a detector with an assumption that all incident power is absorbed in the quantum superlattice. The result showed that the responsivity value becomes comparable to the quantum limit of responsivity:

$$R_{\text{quantum}} = \frac{e}{\hbar\omega} \quad \text{(quantum efficiency)}, \tag{5.6}$$

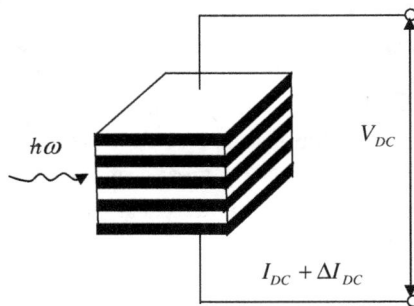

Fig. 5.6. Superlattice based DC detector.

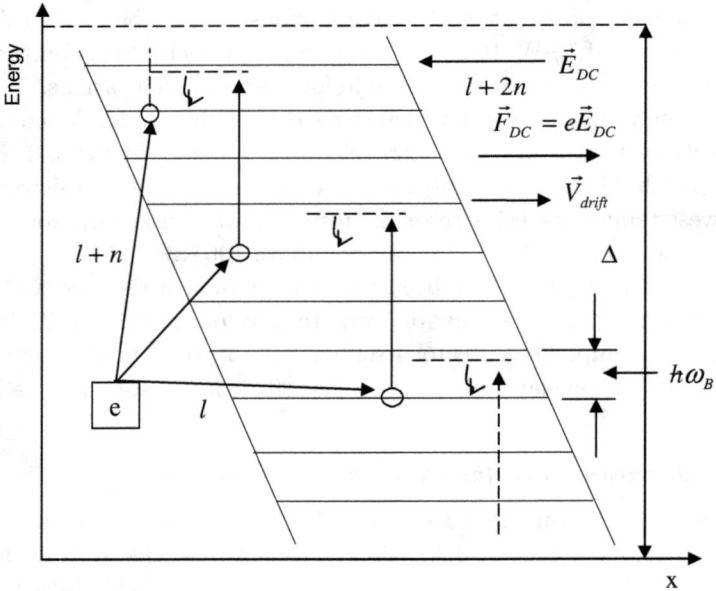

Fig. 5.7.   Electron climbing the Wannier–Stark ladder: Principle of detection.

when the incident radiation frequency ω exceeds the frequency of Bloch oscillations ω_B:

$$\text{where} \quad \omega_B = \frac{eE_d}{\hbar} \tag{5.7}$$

and e is the electron charge, $\hbar$ is the Plank's constant, E is the electric field inside the superlattice. Furthermore, A.A. Ignatov *et al.* did not suggest cooling of the superlattice. At $\omega > \omega_B$, the electrons are climbing up the Wannier–Stark ladder (Fig. 5.7) due to the photon absorption accompanied by photon emission $l \to l+1$ (or $l \to l+n$; $n = 1, 2 \ldots$). By subsequent transitions $l \to l+1$, $l+1 \to l+2$, etc., electrons move against drift velocity. Later A.A. Ignatov and A. P. Jauho developed the self-consistent theory of THz detection by one-dimensional (1-D) layered superlattice coupled to a broadband bow–tie antenna by means of the equivalent transmission line. The conventional design of this detector is shown in Fig. 5.8. In this figure, THz radiation is coupled to an N-period semiconductor superlattice by a coplanar broadband bow–tie antenna. P_i and P_r are the incident and reflected powers, respectively. The equivalent transmission line for this TH-photon detector with a DC bias voltage is shown in Fig. 5.9, where SL is

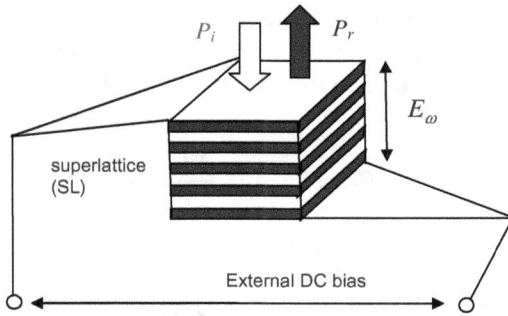

Fig. 5.8.   THz photon detector based on layered superlattice.

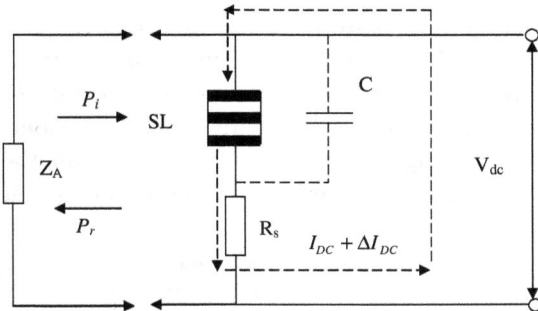

Fig. 5.9.   Equivalent transmission line for the THz-photon detector based on a layered superlattice.

superlattice (miniband electron capable to perform Bloch oscillations), C is superlattice capacitance, R_s is parasitic high-frequency resistance, Z_A is bow–tie antenna impedance, V_{dc} is dc bias voltage, ΔI_{dc} induced DC current change.

The equivalent circuit is similar to that used in resonant tunneling and Schottky diode simulation and allows one treating microscopically the high-frequency response of the miniband electrons and, simultaneously, taking into account a finite matching efficiency between the detector antenna and the superlattice in the presence of parasitic losses.

The current development of THz-photon detector with a quantum superlattice is concentrated on exploiting of 1-D layered superlattice as an active medium instead of a layered superlattice. If the latter is implemented, then the broadband antenna should be changed and a new resonant detector with frequency-selective properties designed.

The current responsivity (compared to Eq. (5.6)) is defined by J.R. Tucker and M. J. Feldman for the superlattice detector as the current change ΔI_{DC} induced in the external DC circuit per incoming AC signal power P_i:

$$R_i(\omega, V_{DC}) = \frac{\Delta I_{DC}}{P_i}. \qquad (5.8)$$

The lateral (surface) superlattice is a superlattice where an additional potential is produced for 2-D electrons localized near a surface of a semiconductor substrate. It can be 1-D chains of identically coupled GaAs/GaAlAs quantum dots (Fig. 5.10) — a form of quantum wires, sometimes called quantum boxes or quantum dots (e.g. GaAs) embedded in a thin epilayer of GaAlAs (Fig. 5.11).

The lateral superlattices are grown by Molecular Beam Epitaxy (MBE) and by Metalorganic Chemical Vapor Deposition (MOSVD). The lateral superlattices can also be grown with an upper substrate (Fig. 5.12). The lateral lattice may be more complex with three-dimensional (3-D) cluster lattices which are a range of lattices grown on one another. The energy spectrum for the 2-D surface superlattice is more complicated than that for

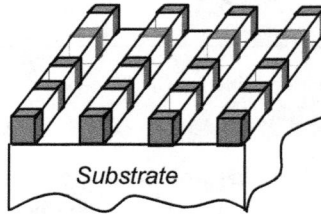

Fig. 5.10.   Lateral superlattice with quantum wires.

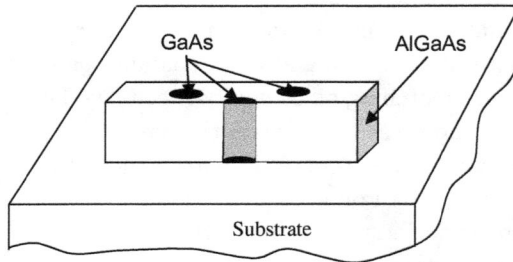

Fig. 5.11.   Lateral superlattice with quantum dots embedded into epilayer.

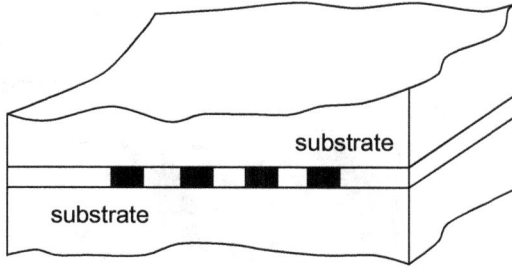

Fig. 5.12.   Lateral superlattice with a substrate grown on the upper side.

the 1-D layered superlattice. Therefore, we shall assume that the internal electric field in the lateral superlattice has only one nonzero component parallel to one of the compositional potential directions so that the energy spectrum is given by:

$$\varepsilon(p) = (\Delta/2)[1 - \cos(pd/\hbar), \tag{5.9}$$

where d is the superlattice period, Δ is the width of the first allowed superlattice miniband; p is the electron momentum parallel to the direction of the superlattice potential.

5.2.1.1. *Example: THz detector based on layered superlattice*

In Fig. 5.13, there is an example of a possible design of the detector based on a lateral quantum superlattice (Raspopin and Cui, 2005). Comparing to the design depicted in Fig. 5.8, we can see that the main distinguishing feature is the change in the method of the attachment of the broadband THz antenna. In this design, both antenna tips lie on the same substrate, on which a lateral (surface) superlattice is grown, being attached to the superlattice by the Ohmic contacts. Raspopin *et al.* created the superlattice on a semiconducting substrate with a grown thin epilayer of *GaAlAs*.

Quantum dots were included in this epilayer as shown in Fig. 5.13. The quantum dots (or they can be quantum wires) are essential for building of the superlattice since they lead to formation of minibands in this quantum structure.

The detector design is independent of scale and dimensions are arbitrary. The authors indicated only two principal dimensions here. One is h which is usually 0–1 μm and the length "a" of the contact between the antenna's tip and superlattice. This dimension depends on the width

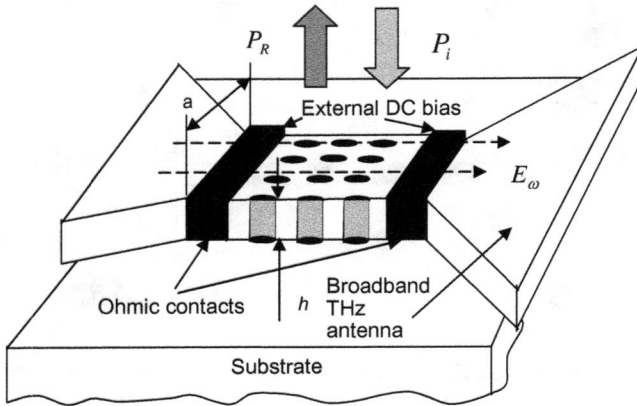

Fig. 5.13.   Proposed detector based on lateral quantum superlattice.

of the grown superlattice and affects the responsivity. The structure may be grown by MBE. In Fig. 5.13, P_i. and P_R are the incident and reflected wave powers, respectively. The external DC bias is applied through the Ohmic contacts as it is shown in Fig. 5.13. Unlike in the conventional superlattice design, the proposed detector's high frequency electric field (E_ω) is parallel to the antenna's plane (rather than being perpendicular). Raspopin *et al.* suggested to switch to the lateral superlattice as a THz detector has several benefits. First one is related to the electron transport comparison. The best value of the effective scattering frequency in the lateral superlattice is $f_{sc}^{min} \approx 0.05$ THz at room temperature. Also, this "plane" detector configuration allows creation of a resonant structure. In this case, the high-frequency current excited in the antenna and flowing through the superlattice pumps the resonator.

5.2.1.2. *Resonant detector based on lateral superlattice*

Figure 5.14 shows a possible design of the proposed frequency-selective THz-photon detector based on a lateral superlattice with a built-in resonator. The principle is based on a standing wave enhancement of the detector responsivity. The detector consists of a lateral superlattice overgrown with an upper substrate, a broadband antenna and a resonator formed by two metal mirrors. The authors indicated MBE as a possible method. The width L must satisfy the resonance condition $L \approx \lambda/2$ (λ is the wavelength of the resonator, $\lambda = c/f\sqrt{\varepsilon_s}$, $c = 3 \times 10^8$ ms^{-1}, ε_s is the THz

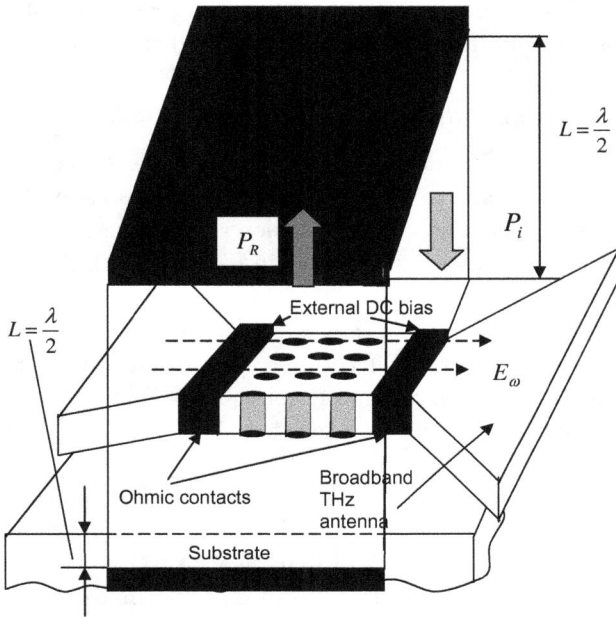

Fig. 5.14.   Metal mirror frequency-selective THz-photon detector.

medium permittivity), for some frequency f_0, near which a responsivity enhancement is desired.

The effect of a resonant enhancement may be achieved in such a detector near this frequency only under the condition that the Q-factor of this resonator is high, which is possible if the characteristic dimensions of the metal mirrors are comparable with λ. In Fig. 5.14, the basic configuration of the resonant detector is given. The authors also suggested modifications to the design. For example, the lower (with respect to the illuminated antenna surface) mirror may not be necessary if the dimension a (see Fig. 5.13) is comparable with λ. In this case, the resonator is formed by the upper mirror and antenna plane. At the same time, the presence of the upper mirror is deemed to be necessary even if $a = \lambda$ because of the skin-effect physics: the high-frequency current excited in antenna is localized near the illuminated antenna surface with a deep-layer and cannot efficiently pump the lower resonator.

It was suggested that the resonator design may be applicable to lower frequency ranges than THz. However, a superlattice is not the only option among semiconductor materials. Since there is only one condition which is: the nonlinear medium must be a lateral (surface) structure.

5.2.1.3. *Responsivity of the resonant detector*

Let us consider the frequency dependence of the responsivity of the proposed frequency-selective detector based on a lateral superlattice with a built-in resonator. The transmission line description is used for this purpose. The equivalent circuit diagram is shown in Fig. 5.15 with the same signs as in Fig. 5.9.

The above detector description does not take into account the skin-effect in the antenna and the diffraction corrections for the electromagnetic field distribution inside the resonator. For this design, the transmission line correctly describes the structure when the dimension a (see Fig. 5.13) is much smaller than λ.

The characteristic theoretical dependence of the dimensionless normalized responsivity (responsivity divided by quantum efficiency) of the proposed resonant detector on the frequency of the incident radiation is shown in Fig. 5.16 (solid line). Also shown is the normalized responsivity of the existing detectors (dash line) with the assumption (for comparison) that both detectors have the same effective high-frequency superlattice impedance and operate at room temperature.

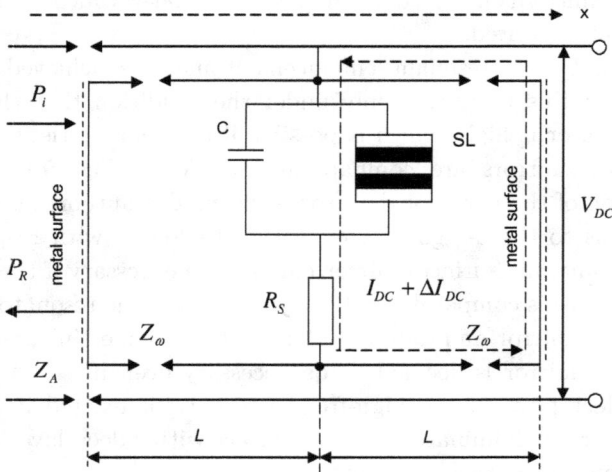

Fig. 5.15. The equivalent transmission line for the frequency-selective THz-photon detector based on lateral superlattice with a built-in resonator.

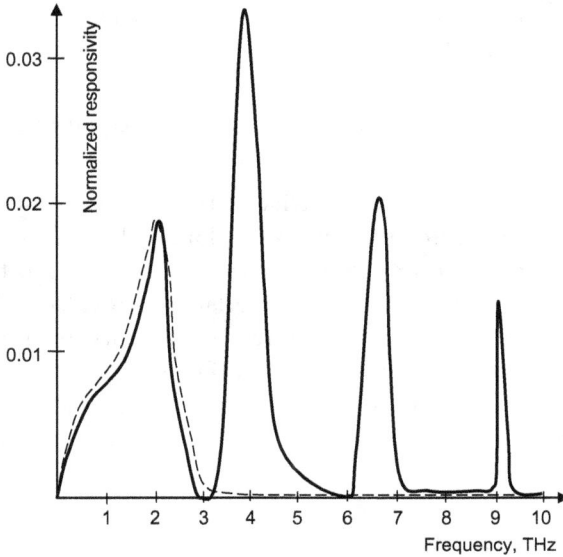

Fig. 5.16.   Frequency dependence of normalized responsivity of the proposed detector.

The following superlattice parameters were chosen: the superlattice period $d = 5\,\mathrm{nm}$; the number of periods $N = 100$; the Ohmic contact areas $S = 9.0\,\mu\mathrm{m}^2$, the maximum current density $j_p = 100\,\mathrm{kA/cm}^2$, the effective scattering frequency $f_\nu = 0.5\,\mathrm{THz}$; the resistance of the parasitic high-frequency losses $R_\mathrm{s} = 0.1\,\mathrm{Ohm}$; the lowest resonance frequency $f_0 = 3.0\,\mathrm{THz}$ and the wave impedance of the broadband antenna $Z_\mathrm{A} = Z_\omega = 104\,\mathrm{Ohm}$, the normalized applied bias $V_\mathrm{DC}/V_P = 0.7$ ($V_P = N_d E_P$, where $E_P = 4\,\mathrm{kV/cm}$), where the electric field corresponds to the maximum superlattice current density.

It can be deduced from Fig. 5.16 that the frequency-selective quality of the proposed detector at one of the multiple resonances, e.g. at $f = 6.1\,\mathrm{THz}$ is $\frac{f}{\delta f} = 440$, and the responsivity of the proposed detector $R_\mathrm{proposed}(6.1\,\mathrm{THz}) = 0.008\,\mathrm{A/W}$. This makes the device useful for the frequency-selective detection in the THz range. Here, the minimal characteristic detection time $t^{exit}_{\min} = \left(\frac{1}{f_\nu}\right) = 2 \times 10^{-12}\,\mathrm{s}$. For the detection with the enhanced responsivity at resonance, e.g. $f = 6.1\,\mathrm{THz}$ by the proposed nanostructure the rise time can be estimated as $t^{\mathrm{proposed}}_{\min} = \left(\frac{1}{\delta f}\right) + \left(\frac{1}{f_\nu}\right) = 7 \times 10^{-11}\,\mathrm{s}$ for chosen structure parameters. For the sake of simplicity Z_ω and Z_A were artificially equated, but as the calculations show, the wave impedance Z_ω influences mainly the shift of responsivity resonance peak near the frequency of the locked resonator (f_0).

5.2.1.4. *Conclusion for example "THz detector based on layered superlattice"*

In the above example, the responsivity of the superlattice-based detector was obtained. It is approximately several hundred times higher than for the existent structures near the resonance frequency due to the matching between the incident wave and the superlattice provided by the built-in resonator. On the other hand, selective responsivity restricts the scope of possible applications for the detector. The estimated rise time of the detector is approximately 10^{-11} s. It is advantageous that the device operates at room temperature.

5.3. Processing electronics

Processing electronics mostly serve the purpose of making the signal, which is received from the detector, suitable for the analyst or an automatic identification system. A number of different devices and systems are employed to process the signal. They include (but are not restricted to) down and up converters/frequency multipliers, local oscillators (LO), couplers, amplifiers, frequency synthesizers and some others. We will consider only some of them, the ones that are more pertinent to the specifics of THz signal processing.

Down-converters: In the traditional system, the channelization and down-conversion of the RF signal are performed entirely in the electronic domain: channelization is effected by a bank of electrical filters, while down-conversion is effected by mixing of the channelized RF signals with signals from electronic oscillators (LOs). If this system was required to be capable of handling hundreds of channels simultaneously, then the size, weight, and power demand could be so large as to render the system impractical. Therefore, in practice, it is common to provide for switching a group of fewer channels onto a channelizer at a given time.

Up-converters multiply frequencies in millimeter range to THz frequencies. Up-converters employ high performance GaAs Schottky beam-lead diodes and balanced configuration with a moderate LO pumping level. It is also possible to perform up-conversion using a passively mode-locked multisection distributed-feedback laser under the external optical injection.

Currently, there are commercially available turnkey up-converters (e.g. "Ducommun Technologies).

An example of processing electronics for THz range, a channel of a field transistor may act as a resonant cavity for the plasma waves. For micron or sub-micron gate length, the fundamental frequency of the cavity is in the THz range. This device can be used for the resonance detection and mixing of electromagnetic radiation of the THz frequencies. The principle of operation is based on the fact that under certain conditions, the steady state with a DC current in the field effect transistor is unstable against spontaneous generation of plasma waves. This instability leads to generation of electromagnetic radiation. The plasma instability can, however, be suppressed if there are appreciable losses at the contacts (Dyakonov and Shur, 1993).

5.3.1. *Example: THz source/frequency multiplier*

In this example, we will consider an LO that is used in down- and up-converters and is part of processing electronics. Example presents a compact THz LO source that was developed by Virginia Diodes in collaboration with the Space Research Organization, Netherlands. The source uses GaAs Schottky barrier diodes to frequency-multiply the power input from a millimeter wave amplifier. The final element in the multiplier chain is a frequency tripler to the WR-0.65 waveguide band, spanning from 1.1–1.7 THz. The tripler generates approximately 10 microwatts of power if pumped with 3 mW (Hesler *et al.*, 2005).

5.3.1.1. *The active multiplier chain*

Virginia Diodes has developed a series of broadband frequency multipliers based on integrated diode circuits. Their tripler to the WR-0.65 waveguide band reportedly requires milliwatts of input power anywhere in the range of 367–567 GHz to generate about 10 microwatts in the 1.1–1.7 THz range. A complete source based on an input signal from a standard low frequency (<256 GHz) synthesizer and generating 5–20 microwatts across a 100 GHz electronic tuning band has been demonstrated. The active multiplier chain (× 72) consists of an input doubler/amp with cooling fan (Space Labs), an integrated (×2 × 2) frequency quadrupler and a cascade of two frequency triplers. The length of the chain is six inches and has no mechanical tuners. The authors used an integrated diagonal horn as an antenna that emits the THz radiation. A primary component of the Virginia Diodes frequency

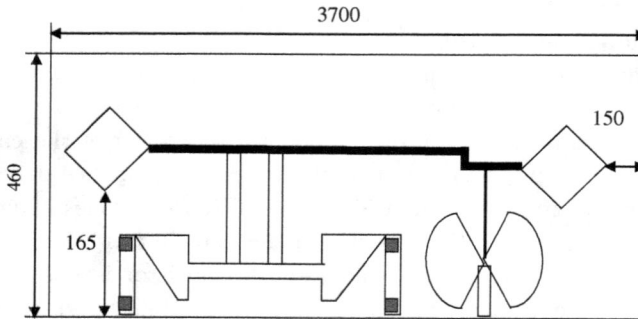

Fig. 5.17.   An integrated GaAs-on-quartz diode circuit mounted on a waveguide housing. The diode bias controls the diode capacitance, thereby controlling the phase of the transmitted signal. 180° phase shift is achieved at 220 GHz.

multipliers is the integrated diode circuit given in Fig. 5.17. The figure shows an integrated GaAs-on-quartz diode circuit mounted onto waveguide housing. The quartz circuit consists of two waveguide probes, frequency filters, impedance matching elements, other passive circuit elements and the GaAs diode. The GaAs is several microns thick and serves in the areas required as the Ohmic and Schottky contacts. The critical diode regions are defined by photolithography and unnecessary GaAs are removed. The integration process minimizes shunt capacitance, achieves alignment precision at the micron level thus eliminating the need for handling, aligning and soldering the diode chips.

The output power of the frequency quadrupler is shown in Fig. 5.18. This component is a cascade of two varactor doublers mounted into a single housing. The bias voltage on each doubler can be adjusted to achieve maximum tuning band. The total efficiency of the quadrupler peaks at about 10%. The bias of the two doubler stages can be adjusted to achieve maximum bandwidth, as shown in Fig. 5.18. However, similar performance is achieved with fixed voltage bias over most of the frequency range.

The output power of the first frequency tripler is shown in Fig. 5.19. This component operates across the entire WR-2.2 waveguide band. Its typical efficiency is 3–5% depending on the output power level and frequency. Significant ripple is seen in Fig. 5.19, particularly at the lower edge of the band. The ripple pattern is caused by standing wave effect. When driven with quarter power levels, several milliwatts of output power are achieved. So far this has been fairly common in cascaded multiplier chains and is generally caused by standing waves between the nonlinear

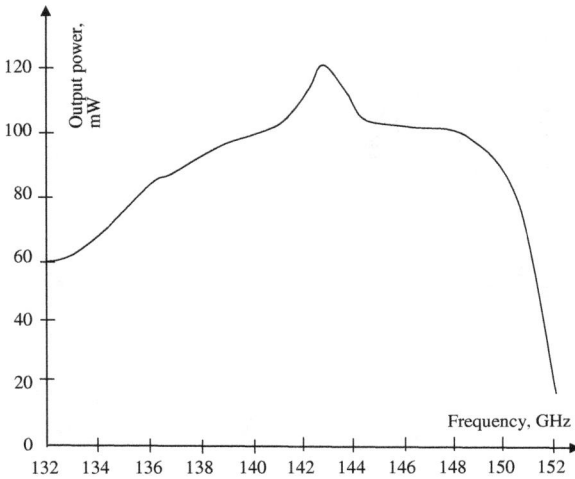

Fig. 5.18. The output power from the Q145 frequency quadrupler.

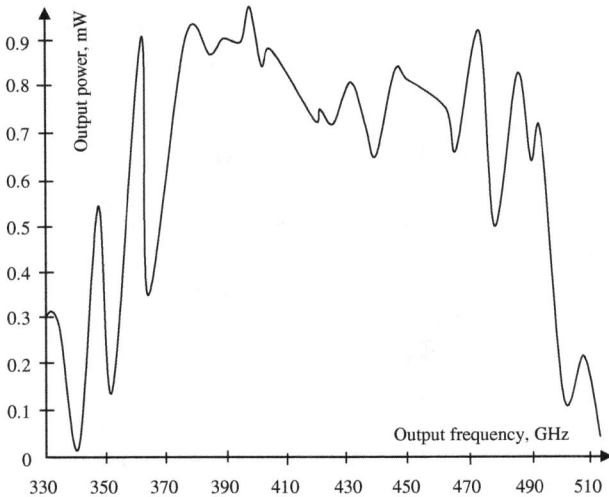

Fig. 5.19. The measured output power of the WR 2.2 × 3 frequency tripler when driven with about 25 mW.

components. Good impedance matching can reduce this effect to negligible levels. However, this is often difficult to achieve in the entire waveguide band, especially since the input impedance of each component will change with the input power level.

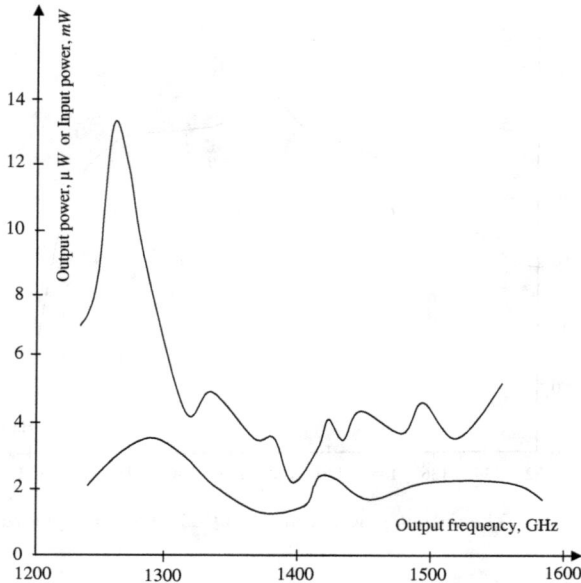

Fig. 5.20.   The measured output power from the active multiplier chain. The authors used two different driver modules, consisting of the input doubler/amplifier and the quadrupler.

The output power of the complete active multiplier chain is shown in Fig. 5.20. The input power to the final tripler is also shown in Fig. 5.20. The first driver generates about 2–3 mW input to the tripler and produces an 8–12 microwatt output from 1250–1320 GHz. The second driver has a lower power of 0.5–2 mW but wider bandwidth. It generates about 2–4 microwatts from 1350–1535 GHz. Other driver modules can be used to achieve coverage over the frequency band from 1.1–1.7 THz and the power of 20 microwatts was achieved near 1.3 THz using a higher power driver. An important future goal would be to make individual drivers that would cover the entire frequency range with sufficient power.

5.3.1.2. *Terahertz source evaluation*

The authors tested one of the active multiplier chains with the goal of demonstrating its suitability for the use as a local oscillator for HEB mixers. The test gas contained ethanol with some water vapor. Although the power level has already been shown to be sufficient by direct measurements, other requirements include narrow line width, absence of spurious signals,

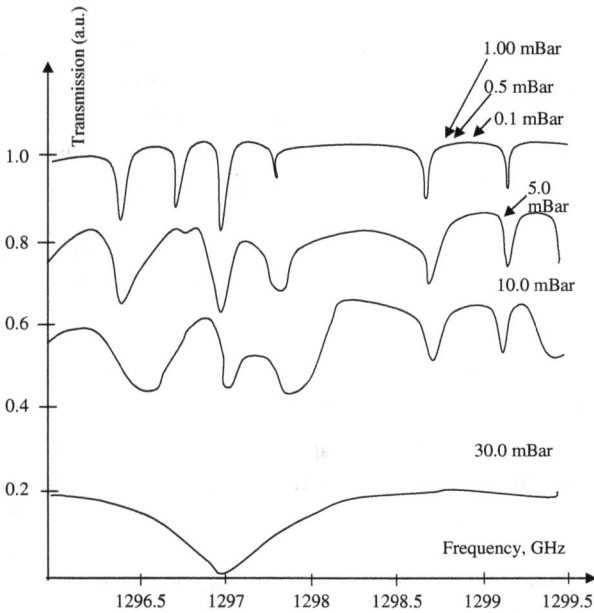

Fig. 5.21. A measurement of the absorption lines of ethanol and water at various pressures in a one meter long gas cell.

stability and low noise. In these measurements, the active multiplier chain was driven with a standard Rhode & Schwarz synthesizer (SMP). The output (THz frequency) was coupled to the bolometer with a beam splitter and the LO coupling loss was about 13 dB. The coupled power was adjusted with a rotating polarizer.

Spectral analysis by the Fourier Transform Spectroscopy indicated that the output signal was pure with unwanted harmonics and other frequencies at least 20 dB below the main signal. Figure 5.21 shows a spectral measurement of ethanol gas (with a small amount of water). This result indicates the narrow line width of the source and the lack of substantial spurious frequency interference.

5.4. Imaging using THz radiation

High-resolution imaging for THz identification has become available with the advent of practical sources and detectors as well as processing electronics capable of satisfactorily dealing with the frequencies in the THz range. There are a number of aspects that are critical to image acquisition. We will consider them in the context of concrete techniques used for THz imaging. One of the most widespread methods at this time uses electro-optical or photoconductive antenna techniques to generate and record broadband pulsed signals. Research into the other optical methods (e.g. mixing two continuous wave (CW) laser frequencies) is ongoing, as is the development of all-electronic THz transceivers. First, we will discuss imaging considerations and major parameters using examples from research and industrial projects, and then safety measures pertaining to THz radiation (concerning imaging systems) are presented.

5.4.1. *Imaging considerations, measurement time and pulse signal-to-noise ratio*

In order to utilize the available spatial and spectral information, large amounts of data must be collected in the spatial and temporal domains. The volume of data depends on the desired resolutions in the temporal and spatial domains, hardware characteristics and many accompanying circumstantial factors. Among them, one rule imposes the lowest limit on the data size. It is the Nyquist criterion that establishes the minimum number of time data points, M_t and pixels, M_{xy} to be acquired. According to this criterion, the sampling interval must not be less than double the highest frequency to be detected in time or spatial domains. If we assume that the spot in focus has a diameter approximately equal to the diameter of the Airy disk, then the pixel size should be at least 4.6 times smaller than the size of the disk.

For example, if the Airy disk diameter, i.e. the diameter of the THz focal spot is 1 mm then the sampling step must not exceed 200 μm. If the image square is limited to 20×20 mm^2, then the minimum number of points that must be acquired to avoid aliasing and undersampling, and to reduce the required spatial frequencies is $M_{xy} = 128 \times 128$ (rounded to the nearest power of 2).

The above considerations are valid only for the X–Y dimension of the data array.

However, similar arguments may be applied for calculations of the minimum number of points required in the time domain to adequately represent spatial responses. This number depends on the THz frequency bandwidth and the desired spectral resolution. If a conventional $ZnTe$ emitter is used, then the highest frequency is in the range of 3–5 THz. This highest detected frequency determines the sampling interval in the time domain, $\delta t \leq 1/(2f_{\max})$, which for this case is near to 100 fs. For a spatial resolution of $\Delta f \approx 50\,\text{Hz}$, the number of points required in the time domain, $M_t = 1/(\delta t \Delta f)$ is also close to 128. However, for many cases this spatial resolution may not be sufficient and then the number of points required in the time domain could be as high as 1,024. It is important to emphasize that THZ pulsed identification (TPI) offers an advantage over many existing imaging techniques because it offers spatial localization, combined with spectroscopic capabilities in the wide range of frequencies: from DC to several THz.

Combining the spatial and temporal dimensions gives a total number of acquisition points approximately 10^6–10^7. Using mechanical scanning systems, the data collection of a single spectral image can take many hours. Notwithstanding this drawback, these systems are still widely used in THz experiments since their modulation and gating methods enhance the SNR of a measured pulse. At present, the SNR values can be about 10^6 for these types of systems.

In order to reduce acquisition time, the properties restricting the bandwidth, should be sacrificed (the application of narrow band THz emitters working at fixed frequencies), or we need to achieve simultaneous acquisition in two of the three dimensions (the application of multi-element detector arrays). The usage of arrays, however, reduces the SNR, thus becoming the main sources of noise fluctuations in the ultrafast femtosecond laser and Johnson noise in detectors and resulting in a subsequent degradation in image quality. Below, we consider several solutions to reduce acquisition time.

In particular, Mittleman *et al.* (1996) discussed the implementation of a fast scanning optical delay line in which the delay stage continuously oscillates at a frequency of 100 Hz. This approach could reduce acquisition times for individual pixels from several minutes to $25\,\mu\text{s}$, with a subsequent reduction in signal-to-noise ratio due to the loss of noise filtering capabilities. Jiang and Zhang (1999) presented a method using broad beam illumination and a CCD system of measurement that could acquire an entire array of pixels at a single time point, in roughly 25 ms. The SNR for these systems is approximately 1×10^3. Later, Jiang and Zhang (2001) developed an alternative method to alleviate the need for mechanical stages, using a linearly chirped laser pulse. According to their statement, this method required only milliseconds to record the full time-domain data for a single pulse. The downside of their method is deterioration of spectral resolution due to the limitations of measurement duration and temporal resolution. As a result, the SNR is only about 60. Of course, averaging can be used to improve the SNR of these faster acquisition methods, but then the gain in speed suffers. To achieve performance comparable with slower systems, it is necessary that the averaging time is usually comparable to the acquisition time of the slower systems (Jiang and Zhang, 1999).

5.4.2. *Parametric images*

From the abundance of data collected for each pixel a large number of parameters can be used to form THz images. Pulse height and shape, and delays in the time domain provide contrast information. Additionally, the spectral content of the pulse can be exploited for pixel-by-pixel spectroscopic analysis and material characterization, including transmittance and absorbance parameters, as commonly used in the Fourier transform infrared spectroscopy. The above parameters may be combined. Combinations of these parameters are useful for developing thickness-independent parameters that allow differentiation between time delays due to refractive index effects or sample thickness. In many cases, the parameters can be compared to a reference pulse to receive their characteristics. Similar parameters can be used for both reflection and transmission images.

Some commonly used image parameters are described below. These are separated into two main categories: those based on the pulse profile in the time domain and frequency domain parameters, determined using information at individual frequencies or combinations of frequencies in the spectral domain.

Time-domain parameters:

- **Pulse amplitude.** Maximum value of pulse in time domain for each pixel. It can be expressed relative to the reference amplitude and represents the proportion of transmitted signal for each pixel.
- **Pulse time delay.** Time difference between the maximum of reference and pixel pulses in time domain. The thickness and refractive index of a material determine this delay.
- **Pulse width.** Full width at half maximum of pixel pulse. It depends on the ratio of low- and high-frequency components and an increase (pulse broadening) implies preferential absorption of high frequencies.
- **Instantaneous electric field.** The electric field amplitude imaged at points in the time domain. This shows the propagation of the electric field through the material.

Frequency domain parameters:

- **Transmittance.** Ratio of transmitted-to-incident intensities at a given frequency, determined by the sample thickness and absorption coefficient.

The use of integrals over bands in the spectrum can improve noise characteristics for this and other frequency domain parameters.

- **Time delay.** Delay time due to the presence of the sample, calculated from the phase component of the Fourier transform. It is proportional to the thickness and refractive index of the sample.
- **Absorbance.** Logarithm of the inverse of transmittance.
- **Dual frequency or panchromatic.** Use of two or more frequency components to image (for example, ratio of transmittance at ν_1 relative to transmittance at ν_2).
- **Phase.** Unwrapped phase component from the Fourier transform at the given frequency. This is proportional to the time delay and depends on sample thickness and THz frequency. Commentary: The notions of instantaneous phase and instantaneous frequency are important in signal processing for the representation and analysis of time-varying functions. The instantaneous phase $\phi(t) = \arg(x(t))$. When $\phi(t)$ is constrained to an interval $[-\pi; \pi]$ or $[0; 2\pi]$, it is called *wrapped*. Otherwise, it is called *unwrapped*.
- **Dispersion.** Derivative of transmittance with respect to frequency. This displays the change of absorption as a function of frequency and highlights characteristic peaks in the spectral domain.
- $\alpha(\nu)/(n(\nu) - 1)$. Independent of thickness due to cancellation of the thickness terms by the ratio of the expressions of absorption coefficient and refractive index. It is material dependent.

Figure 5.22 shows an example of pixel pulses transmitted through various thicknesses of a step wedge. The wedge was manufactured from duraform by the selective laser sintering process. From Fig. 5.22, it can be seen that thicker steps attenuate and delay the pulse by greater amounts. This provides a contrast in the parametric images for pulse amplitude and pulse delay in the time domain and transmittance and time delay at 1 THz in frequency domain. The dual frequency transmittance parametric image reveals a contrast between the steps because the higher frequency (1 THz) is more strongly attenuated than the lower (0.8 THz), as is evident in the spectrum shown in Fig. 5.22. By the instantaneous electric field parameter, it is possible to selectively image individual steps by selection of the appropriate time point.

5.4.3. *Image spatial resolution capabilities*

According to the Abbe criterion, the minimum size of a recognizable feature in an image formed using EM radiation is ultimately limited by

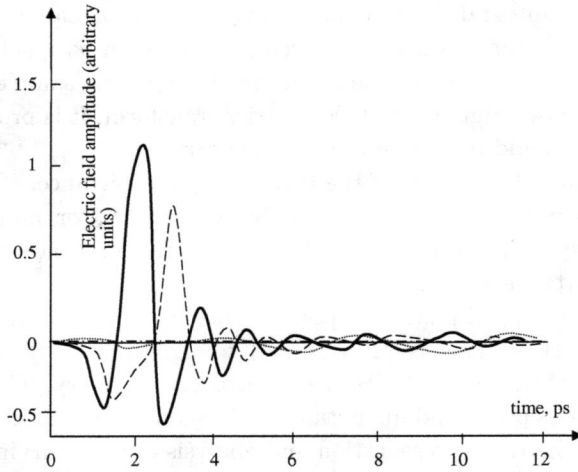

Fig. 5.22. (a) Reference THz pulse (solid line) compared to pulses delayed and attenuated by 0.5 mm (chain curve), 2 mm (broken line) and 4 mm (dotted line) of duraform.

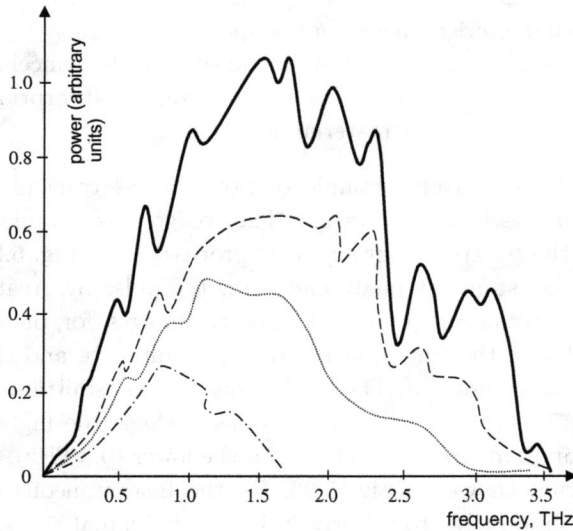

Fig. 5.22. (b) Corresponding spectral content of each pulse.

the wavelength, due to diffraction effects. In practice, however, system properties may be adjusted to avoid the above limitations. Since the THz region of the spectrum could stretch from tenths of a THz to 10 THz and wider, the spatial resolution in the images strongly depends on the frequency range used for imaging.

For example, at 1 THz, assuming minimal scattering, the diffraction-limited beam waist corresponds to a size close to 300 μm, compared to 600 μm at 0.5 THz. An effective method of representing the spatial resolution capabilities of an imaging system is the modulation transfer function (MTF), which represents the modulation of the signal by the imaging system at each spatial frequency.

Due to smaller diffraction limits, higher THz frequency components reproduce higher spatial frequencies better than lower THz frequency components and hence yield a higher MTF. This is illustrated by the example MTF in Fig. 5.23 for a typical THz imaging system.

The MTF curves were determined by measuring the modulation effect on the relative transmittance of the THz signals by a series of deposited gold bands on the TPX (Mitsui, UK), a THz transparent plastic. The arrangement and spatial frequencies of these gold bands are shown in

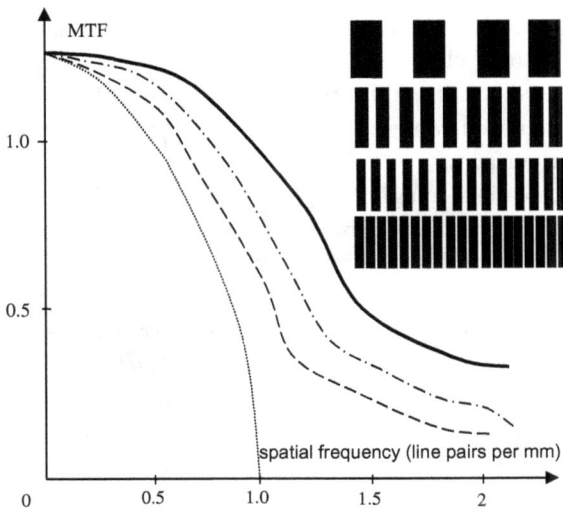

Fig. 5.23. NTF curves determined from relative transmittance parametric images of the gold on TPX test object (inset) at frequencies of 0.5 THz (dotted line), 1.0 THz (broken line), 1.5 THz (chain line) and 2.0 THz (solid line).

the inset (right upper corner) in Fig. 5.23. The improvement of spatial resolution with higher THz frequencies is clearly shown in Fig. 5.23.

Gold bars of 100 nm thickness were deposited on the TPX at the spatial frequencies of 0.25 (top row of inset), 0.50, 0.75 (second row), 1.0, 1.375 (third row), 1.625 and 2.0 line pairs per mm (bottom row).

For THz systems using standard electro-optical techniques, the spectrum above several THz has low intensity, so that the improved resolution is offset by a reduced image SNR at higher frequencies.

Using the principle of confocal microscopy, the size limit of feature recognition may be reduced to $\approx \lambda/\sqrt{2}$. Using a near-field condition, it can be improved even further (Hensche *et al.*, 1997), lately with the spatial resolution of up to $\lambda/1000$ (Chen *et al.*, 2003). An aperture smaller than the diffraction limited beam waist is placed in front of a THz source and the sample is placed immediately beyond this, within the distance comparable to the wavelength so that the near-field conditions are maintained. Here, the improved resolution is again a cause of a large decrease in the signal-to-noise ratio because the aperture strongly attenuates the signal intensity with a radial dependence. The resulting THz pulse goes through a high pass filter so that the spectral content of the transmitted pulse is blue-shifted compared to the pulse incident on the aperture. Wyune *et al.* (1999) quote improvements in spatial resolution of better than a quarter of the wavelength for wavelengths around 1 mm and Hunsche *et al.* (1998) quote values close to 70 μm, determined by using a line spread function analysis across a knife blade. Chen *et al.* (2000) used a dynamic aperture and achieved values of 50 μm, however, this approach is limited to semiconductor surfaces.

Another near-field imaging implementation for THz imaging improvement is the usage of extremely brilliant sources to compensate for intensity losses (to confine the THz radiation at the cost of total power). Among THz sources showing promise is the highly brilliant coherent synchrotron radiation (CSR) from an electron storage ring (Schade *et al.*, 2005). The CSR source is Berliner Elektronen Speichering-Geselschaft fur Synchrotronstrahlung (BESSY). It is a pulsed source with a frequency of 1.25 MHz determined by the time the relativistic electron bunch needs to travel one orbit with a circumference of 240 m. The emitted CSR power varies as a square of decaying detection ring current.

The far-infrared port of the beamline provides a collimated CSR THz beam, which is about 98% linearly polarized. The THz beam passes through a Martin–Puplett spectrometer before being transferred to the Scanning

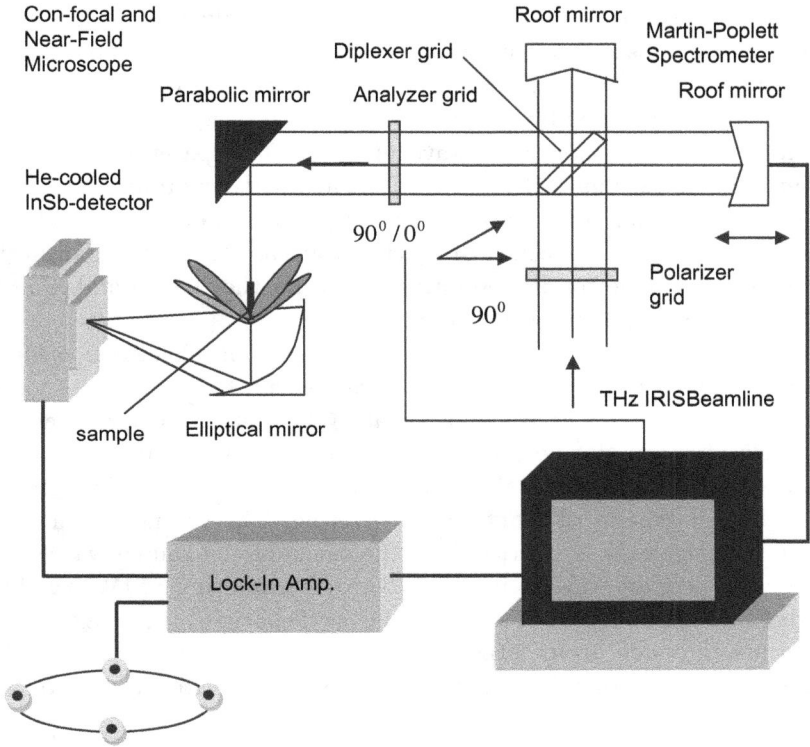

Fig. 5.24. Schematic diagram of the THz scanning near-field infrared microscopy (SNIM) set-up.

Near-field Infrared Microscope (SNIM) (see Fig. 5.24) where it is focused into a conical waveguide with a circular cross-section and an exit aperture of a diameter smaller than the wavelength. The sample is held in front of the exit aperture by a spring providing a direct contact of the sample with the probe. Imaging is performed by moving the sample in front of the exit aperture by means of a computer controlled x–y stage.

The evanescent field at the exit aperture penetrates the sample and the scattered radiation containing the spectral information is collected by an ellipsoidal mirror and is then focused into a LHe-cooled InSb detector. The revolution frequency of the electron bunches is used as a reference for look-in detection of the SNIM signal by a fast liquid He-cooled InSb detector. This detection provides dynamical range of nine orders of magnitude, which is essential for near-field imaging in strongly absorbing samples since it discriminates intensity from the source against thermal background radiation emitted by the beamline, the sample itself and the environment. The image is generated by comparison of the SNIM signal with the sample position relative to the conical waveguide axis. The spatial resolution was estimated to be about $130\,\mu m$ for $200\,\mu m$ aperture probe and can be improved to $70\,\mu m$ using $100\,\mu m$ aperture probe. Taking broadband near-field images, the spatial center of gravity at around $12\,cm^{-1}$ is transmitted yielding an average spatial resolution of $\lambda/6$ for the $200\,\mu m$ and $\lambda/12$ for the $100\,\mu m$-aperture probe respectively.

The THz SNIM concept (primarily meant for biological, medical/ forensics applications) was tested on leaves, where the contrast was mostly formed by the amount of liquid (water) present. Recently, images of living leaves have also been obtained by other authors (Mittelman, 2003; Koch, 2003) from a confocal setup applying THz-TDS to investigate the re-hydration process of plants after watering. The spatial resolution of these investigations is restricted by the range of the source wavelength. However, there are ways to increase the resolution using SNIM technique. Figure 5.25 shows a part of a freshly cut parthenocissus leaf imaged in the transmission mode. In its THz near-field image, an inner structure of the veins is apparent that is mainly formed by liquid water absorption and

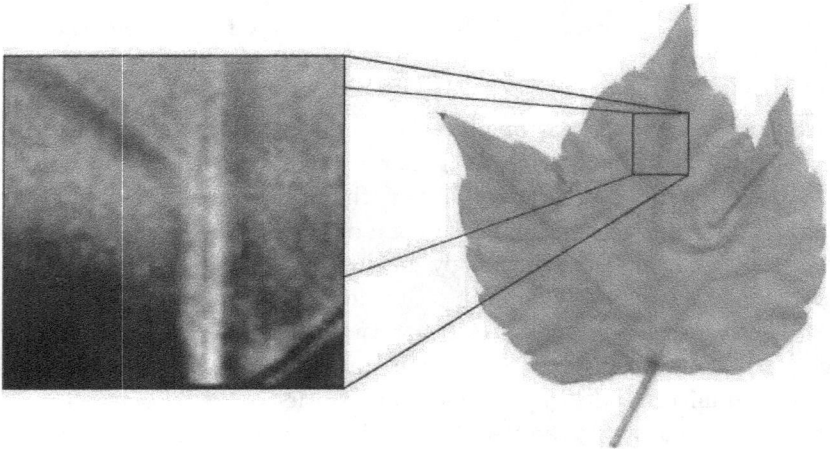

Fig. 5.25.   Near-field image of a section of a parthenocissus leaf. Less absorption is indicated by a darker region in the THz image.

possible scattering at the structural boundaries. Both the THz image and the visible light image may reveal similar object features but the THz SNIM enables studies of hydration dynamics with a high spatial resolution because of the method's sensitivity to the water concentration in the sample. This feature, in particular, is valuable for forensics applications where the degree of dehydration is connected with the time that has elapsed since the tissue was dead.

An alternative approach to near-field imaging is to use a detector to measure the evanescent field emanating from the sample (Mitrofanov *et al.*, 2001). To compensate for strong signal attenuation, the near-field receiving device is optimized by the integration of the GaAs photoconducting detector and the aperture in very close proximity so that the near-field signal can be detected with sufficient sensitivity. Using this technique, the spatial resolution is independent of THz wavelength and is scaled with aperture size. Spatial resolution (based on the 10–90% change in transmittance for a second gold edge) of $39\,\mu$m for a $30\,\mu$m aperture (Mitrofanov *et al.*, 2000) and $7\,\mu$m for a $5\,\mu$m aperture have been demonstrated. The limit of aperture dimensions, given the present system powers and noise levels, is about $3\text{--}4\,\mu$m due to the d^3 attenuation in signal by the aperture of diameter d. The signal, in this case, may be affected in time domain by multiple reflections that occur between the gold screen layer of the detector and the back surface of the sample.

For many *in vivo* applications, reflection geometry will be required to form images of the structure below the surface of the body. The time delays of reflected pulses can be used to calculate the depth of interfaces, in a manner similar to ultrasonic imaging. An extension of this technique using multiple receivers and a Kirchhoff migration image reconstruction technique, originally developed for geophysical prospecting, enables to localize extended objects that are not parallel to the surface (Dorney *et al.*, 2001). The vertical resolution of both these methods is limited by the bandwidth of the incident THz pulse since it follows from the Rayleigh criterion that thin layers cannot be resolved if the separation is much less than the coherence length of the light pulse. For typical THz bandwidths, the limit of depth resolution is close to $100\,\mu$m. Depth resolution can be improved beyond this limit by using interferometric techniques, in analogy to optical coherence tomography (Johnson *et al.*, 2001). At the present moment, the application of a Michelson interferometer and exploiting the Gouy phase shift arising from the lens in the sample arm, depth resolutions close to $10\,\mu$m could be achieved.

5.4.4. *Safety measures for THz radiation*

THz imaging systems used *in vivo* should comply with existing safety guidelines for exposure to electromagnetic radiation. The guidelines for pulsed EM radiation are given in terms of the maximum permissible exposure (MPE). This is a measure of energy relating to the entire radiation train, calculated using root mean square and peak electric and magnetic fields, or plane wave power densities, depending on the frequency range (American National Standard Institute, 2000). Several experimental factors affect the MPE value and must be considered in the safety analysis of a system. These include the pulse repetition frequency, F; the total exposure duration of a region of skin, T; the total number of pulses in time, T, N; the cross-sectional area of the incident beam, A; the time duration of a single pulse t has been applied for a typical compact TPI system using electro-optical or photoconductive antenna sources. It was shown that the limiting average power per pulse, MPE_{pw} is given as:

$$MPE_{pw} = \frac{A \times MPE_{cw}}{Ft}, \tag{5.10}$$

where MPE_{pw} is the continuous exposure limit of $1000\,Wm^{-2}$ for an exposure area under $0.01\,m^2$ (American National Standards Institute, 2000). Typical values of the parameters were used ($F = 82\,MHz$; $t = 10^{-13}$; $A = 7.8 \times 10^{-7}\,m^2$ (resulting from a circular beam of 1 mm diameter; $T = 600\,s$) to give a calculated $MPE_{pw} = 94\,W$. Presently, the average powers of the THz imaging systems using electro-optical or photoconductive generation are below $1\,\mu W$. If amplification is used, the above value can be increased to $1\,mW$ (Cole *et al.*, 2001; Loffler *et al.*, 2000). Hence, current systems have a power output per pulse which lies well below the limit set by the guidelines.

It should be noted, however, that the safety guidelines on which the analysis was based, were derived from experiments performed at shorter wavelengths (less than $10.6\,\mu m$) and using longer pulse durations (over $1.4\,ns$) than those used in TPI.

There are far stronger absorption peaks for the water molecule in the THz region than at shorter wavelength, which could lead to more extensive tissue damage. Similarly, the ultrashort pulses used in the TPI have the potential for thermo-mechanical effects that would not have occurred with nano-second pulses.

5.5. Conclusions

At present, the THz technology is already not a novelty. There are commercially available THz sources, detectors, processing electronics, imaging equipment, etc. Although prices are still high (a detector may cost some 4,000–5,000 US dollars and a source (a laser, etc.) some 30,000–50,000 US dollars), they are already within reasonable limits for military and special purposes. However, the practical THz technology is still rather exotic and more often than not belongs to the laboratory rather than to the field deployment. Although, the sensitivity is often sufficient, in many cases, the time for image processing is still long, not to mention the bulky equipment. In this chapter, we tried to present new and promising ideas that may help to overcome the above-mentioned drawbacks. For example, among the sources, the QCL is one of the popular choices now with its several merits of being able to work in the CW and pulse-mode regime and being capable of delivering 10s of mW at frequencies up to 10 THz. Optically-pumped THz lasers are used quite often as well. In terms of power (one of the critical parameters for standoff mode and deep penetration), they are the ones that top the charts.

The THz detectors seem to present fewer problems than some other THz components: even at the present moment, they are quite compact and comparatively inexpensive. With respect to sensitivity, as one of the principal parameters, quantum superlattices may be a promising direction for the new developments. These developments are also important for THz imaging as well as for the processing electronics. The latter being one of the "bottle necks" for on-site identification systems still being very slow if high resolution needs to be achieved.

Summarizing, we can conclude that a number of attempts have been made towards ubiquitous, mobile, and reliable special and military THz applications, however, the many components and aspects that are involved complicate the design process and slow down the overall progress.

References

Biedron, S, JW Lewellen, SV Milton, N Gopalsami, JF Schneider, L Skutal, Y Li, M Virgo, GP Gallerano, A Doria, E Giovenale, G Messina and IP Spassovsky (2007). Compact, high-power electron beam based terahertz sources. *Proc. IEEE*, 95(8), 1666–1678.

Brown, E (2003). THz generation by photomixing in ultrafast photoconductors. *International Journal for High Speed Electronics*, 13, 497–545.

Chassagnaeux, Y, R Colombelli, W Maineult, S Barbieri, HE Beere, DA Ritchie, SP Khanna, EH Linfield and AG Davies (2009). Electrically pumped photonic-crystal terahertz lasers controlled by boundary conditions. *Nature*, 457, 174–178.

Chen, HT, R Kersting and GC Cho (2003). Terahertz imaging with nanometer resolution. *Appl. Phys. Lett.*, 83, 3009–3011.

Dyakonov, M and MS Shur (1993). *Physics Review Letters*, 71, 2465.

Dunbar, LA, *et al.* (2005). Design, fabrication and optical characterization of quantum cascade lasers at terahertz frequencies using photonic crystal reflectors. *Opt. Express*, 13, 8960–8968.

Hesler, J, D Porterfield, W Bishop, T Crowe, A Baryshev, R Hesper and J Baselmans (2005). Development and characterization of an easy-to-use THz source. *16th International Symposium on Space Terahertz Technology*, Sweden.

Koch, M (2003). Bio-medical applications of THz imaging. In *Sensing with Terahertz Radiation*, pp. 295–316. Springer.

McIntosh, K, E Brown, K Nickols, O McMahon, W DiNatale and T Lyszczarz (1996). Terahertz measurements of resonant planar antennas coupled to low-temperature GaAs photomixers. *Applied Physics Letters*, 69(24).

Sakai (ed.), K (2005). *Terahertz Optoelectronics*. Springer.

Mittelman, D (2003). Terahertz imaging. In *Sensing with Terahertz Radiation*, pp. 117–153. Springer.

Raspopin, A and H-L Cui (2005). Frequency-selective sub-mm-W detector based on semiconductor superlattice with a built-in resonator. *Stevens Institute of Technology*.

Schade, U, K Holldack, MC Martin and D Fried (2005). THz near-field imaging of biological tissues employing synchrotron radiation. *Proc. SPIE*, 5725(46).

Chapter 6

Electronics for Portable THz Devices

The portability of the terahertz (THz) devices may be essential in a number of applications. First of all, for military purposes, the possibility of carrying THz equipment is especially important. That includes early warning systems and systems used for military operations. Secondly, security devices capable of detecting weapons, bio products, chemicals, drugs, etc. and mobile enough to be transported from place to place are all largely dependent on cost-effective, high-power — especially hand-held — portable imaging systems. As an example, one of the Army Research Office calls for proposals a couple of years ago could be named. The need for such a development is not surprising; it is reminiscent of the similar tendency toward computer miniaturization that started about three decades ago on a large scale. In many cases, it is impossible or impractical to install or deploy the heavy and bulky equipment that has been the feature of the THz systems so far. That is why portability comes forward in today's THz issues. It is also a major problem for medical and forensic (as part of medical) applications that a compact, cost-effective, high-power apparatus is still at the formative stage of development. One example of a system that is required by military, special and commercial application is a device for three-dimensional (3-D) images to diagnose dermal abnormalities such as distinguishing between healthy and ailing skin (including burns and fractures in phalanges where there is little muscle tissue around the bones). Such a need arose in connection, in particular, with military operations around the world.

At present, a major obstacle that prevents the full exploitation of the THz range is the lack of affordable, compact (portable), high-power THz sources and detectors (receivers). The current research has demonstrated that semiconductor technology can be utilized to develop THz modules that are fairly compact, however their efficiency is low, they are low-power, expensive and their frequency is difficult to change. Many of the current cutting-edge THz devices can be used only in laboratories. Moreover, the present commercially available THz technology is insufficient in power and has remained on macro scale with only optically-pumped THz laser sources approaching output power levels of one Watt and only for certain THz frequencies.

Current THz technology has allowed demonstration of high-resolution 3-D imaging through a variety of common media (such as clothes, wood, semiconductor materials, ceramics, biological tissues, leather, etc.). High-resolution imaging is necessary for a number of special and military applications including security identification and screening of people, cargo, chemical and biological agents, forensics (including medical identification and diagnostics), the ability of penetration of objects (*e.g.* walls), etc.

A major limitation of current THz imaging is the required collection time of the image. Most high-resolution systems require many minutes and even hours to collect a single image. This is usually due to a lack of cost-effective THz devices causing many such systems to implement only one source. It means that this source must be mechanically scanned across the sample to receive an image. Leveraging current technology and fabrication techniques in order to develop a more cost effective and efficient source and detector would facilitate image collection by implementing more sources and detectors per system. Designing new algorithms and software as well as optimizing current algorithms and software for processing 2-D and 3-D images would also contribute to decreasing time of image collection. Moreover, in the realm of medical THz applications, only diagnostics of dermal abnormalities (such as distinguishing between healthy and cancerous skin) are readily available. Recently, means of THz diagnostics of skin ailments (*e.g.* burns), fractures in phalanges (with little muscle tissue around the bones) are developed.

At the same time, owing to the low-power capabilities of current systems and the inherent absorption of THz RF by polar molecules (*e.g.* water), achieving high levels of penetration (greater than 1.5 mm by experimental set-ups) into human tissue has not been very successful. The signal to noise ratios (SNR) required to detect reflection/transmission of the current systems are high because of the only low-power systems available. With the increasing need for technologies that could utilize a complete THz system, it is important to develop an affordable THz RF system that may be one of the major steps toward a compact THz

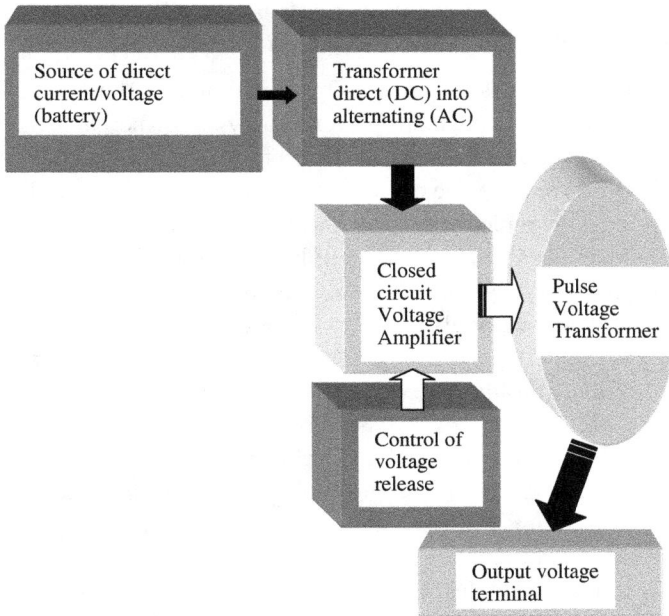

Fig. 6.1.   Power source based on resonance amplification.

system. Not only can such a system be incorporated into many portable applications in the fields of defense security and forensics but it will also make the THz applications more affordable for small businesses (*e.g.* medical diagnostics). Thus, even more expansion in THz research could be realized. There are many possibilities for miniaturization in the field of THz technology: semiconductor and optical elements, power supplies, antennas thus providing opportunities for an introduction of portable (and even hand-held) THz systems in military, special and commercial areas. As far as military and special applications are concerned, the main objectives are to develop a cost effective, portable, high-power THz RF system that can be used to realize battlefield capabilities of THz radiation and aid current Advanced Technology Objectives (ATOs) in their implementation. Presently, the Department of Defense and its agencies aim to design a system with a gross weight less than 20 lbs, the possibility of collection of 2-D and 3-D images, with THz RF transceiver circuit/system. The completed device should be a mature system capable of being deployed in battlefield conditions and providing imaging of military cargo, medical diagnostics of battlefield wounds and personnel security scanning.

6.1. Example: *Resonance amplification* — power source based on resonance amplification

One of the issues that is important for portability is the availability of a compact, light and cost-effective power supply with the efficiency close to 100%. At present, there are reliable and effective chemical power sources (batteries). However the time of their service is restricted, periodically their supply needs to be replenished and their weight may be substantial (*e.g.* in case a soldier needs to carry a supply of batteries). Therefore, a need arose for a power supply that can provide DC voltage for a portable THz system. A number of possible solutions were considered by DARPA and other agencies. One of the possible solutions offered by the author is a resonance voltage amplifier (it was used for one of the DARPA's projects) (Sokolnikov, 2007).

A small 9 V battery was used as a primary source of DC voltage, which is transformed into AC high voltage (Fig. 6.1). Then, the produced voltage would undergo resonance amplification in a closed circuit until it reaches the preset magnitude and is transformed to a higher voltage level by the Pulse-Voltage transformer. The last module is not necessary for every application. Without the module, the voltage at the output of the device is about 100–200 V depending on the design, which corresponds to the range for standard electricity consumers.

References

Sokolnikov, A (2007). THz identification of humans and concealed weapons for law enforcement, government, and commercial applications. *Proc. SPIE*, 6538.

Chapter 7

THz Applications

Originally, terahertz (THz) radiation (also, T-rays) was generated and detected by employing conventional techniques borrowed from microwave and millimeter technologies. Since solid-state Continuous-Wave (CW) and ultrafast pulsed lasers were introduced with biased semiconductors and nonlinear crystals, there have been significant advances in THz technologies. The rapid development of such technologies has allowed shifting their focus to the applications, which are numerous and yet in the formative stage of their development. THz sensing was first, and still is, applied in the areas of astronomy and atmospheric science. As there is an increasing military dependence on space-related systems, it is likely that THz radiation will play a major role in the development of future space capabilities such as satellite-to-satellite communications. Various rotational, vibrational and translational modes of molecules are within the THz range ($\sim$0.1–10 THz). Since these modes are unique for a particular molecule, it is possible to use them for identification of chemical and biological substances, the quality that has encouraged many research efforts, especially for defense, security and forensics. Astronomers and chemists have already utilized THz to characterize and identify many small organic molecules. Their approaches can be applied, for example, in the detection of anthrax and some other infectious diseases providing novel approaches for counter-terrorism. The structure of biomolecules is closely related to their functionality, which provides ways for a wide range of applications in biomedicine, for example, in DNA sensing. On the other hand, THz has low photon energies compared with X-rays, thus it is not harmful to living tissue. In addition, the THz's ability to penetrate a few millimeters of human skin makes the THz detection valuable for detection of skin cancer *in vivo* (*i.e.* of live skin, not a dead tissue). T-rays are also capable of penetration of dielectric materials, something that cannot be done effectively outside this electromagnetic range. This allows detecting chemical and biological agents, and the remote imaging of personnel to detect hidden plastic and metal objects.

Sensing in this Electromagnetic (EM) range potentially provides opportunities in a number of areas of interest towards defense and security, such as short range radar sensing, as THz wave can better penetrate through, *e.g.* fog than visible light. The lower scattering of THz means significantly better imaging in IR spectrum. In medical transillumination applications, much higher contrast can be achieved than with visible light or X-ray. The same EM qualities give better tomographical images of ceramics, as an example. This possibility comes from time-delay detection of reflected THz pulses from the surface of the investigated structure. In addition to their defense, security and forensic applications, the commercial effect can be achieved in industry for quality control with reduced cost for surface defects locating. Although the possible applications of THz phenomena are quite numerous even at the present moment, the practically realizable applications are rather limited even for the high risk special and military purposes where investments tend to be larger than for the sake of scientific and commercial projects. The most substantial areas are detection of concealed weapons, explosives and nonmetallic structures. The medical applications are discussed in a separate chapter. Fancier developments are still at the formative stage (*e.g.* THz cell phones, etc.) and are unlikely choice for military and security purpose, although THz communication shows some promise.

This chapter's discussion involves experimental and industrial examples of THz applications along with the current and potential problems facing the new THz technology.

7.1. THz imaging of nonmetallic structures

The first real-time imaging system using the THz band was introduced in 1995, more than a decade ago (Sokolnikov, 2007a). Since then, numerous applications have been introduced and many problems solved. Some of them were overcome by the introduction of THz time-domain spectroscopy (THz-TDS). This system uses femtosecond pulses of near-visible laser light to opto-electronically generate a coherent THz wave. The resulting electromagnetic pulse is a broadband and spans from lower than 100 GHz to higher than 2 THz. The receiver structure requires the simultaneous arrival of a delayed femtosecond laser pulse and the generated THz wave. Through this arrangement, the laser pulse system provides extremely bright, coherent emissions onto a gated receiver with sensitivity several orders of magnitude higher than most bolometric (thermal) counterparts.

THz imaging permits three-dimensional tomographic imaging of non-metallic objects, which includes the detection of faults, specific structure details of the inner design of nonmetallic parts — the fact that can be useful for identification of a unique password tablet. The possibility exists to combine spectroscopic characterization and/or identification with pixel-by-pixel imaging. As in many applications, the features that we want to detect are subtle but feasible for THz identification since an interaction of a small feature with a single-cycle THz pulse imposes only a small additional distortion on the waveform. A good example is "detecting" a disbanding (rupture) between two surfaces. It is possible to distinguish a gap on the order of the coherence length of the THz pulse. Further, this possibility is implemented to make an identification tablet. There are occasions when identification of individuals in civil clothes or uniform may be necessary. In such a case, the ability of THz radiation to react to different configuration of, for example, plastic objects, such as above-mentioned tablets, etc. may be used to give a warning sign if the approaching subject is foreign, be he or she in uniform (the one that may look like your own uniform) or in civil clothes. The identification gives an advantage of "checking the credentials" without asking for them. This feature may be especially useful when a military detachment is located close to the adversary occupied territory. The idea for such an identification device came from real situations. Below is a detailed description of the physics of the tablet functioning.

7.1.1. *Identification of tablet structure*

In our case, the tablet is the object of an analysis (identification) by a pulsed THz imaging. The tablet's coatings are semi-transparent to the THz frequencies and do not scatter them significantly. The THz pulses, incident on a tablet interface, penetrate through the different coating layers (Fig. 7.1). At each interface, a portion of the THz pulse is reflected back to the detector. The amplitude of the reflected THz radiation is recorded as a function of time. The technique operates much the same way as ultrasound. The sample is completely unaffected by the measurement. The layer thickness uniformity is established simply from the transit time of the pulse to each surface. With knowledge of the refractive index of the layers, the actual thickness can be determined to a depth resolution of approx.

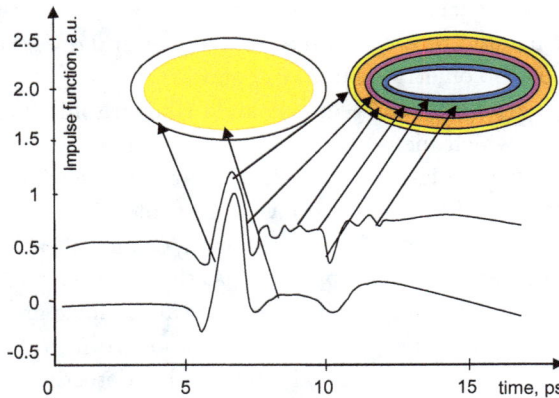

Fig. 7.1.   Traces show the reaction of the inner layers on the incident radiation.

$20\,\mu$. The spot size of the THz pulse, and, therefore, the lateral resolution, is about $250\,\mu$. Below, we shall show the physics behind the imaging as well as the choice of the materials for the tablet.

7.1.2. *Interferometry for Terahertz imaging — a possible solution*

In order to improve the detectability of subtle features, interferometry is used in combination with THz tomography. This idea has analogies in optical coherence tomography, in which the signal pulse, reflected off the sample, is interfered with a reference wave to provide enhanced sensitivity (Sokolnikov, 2005). The general principle is illustrated in Fig. 7.2.

The pulse is inverted in the reference arm. At a known sample depth, the main pulse destructively interferes with the inverted pulse. Any sample that affects the THz signal causes the cancellation in the reference arm. The collimated THz beam is directed into a Michelson interferometer (similar to the one in Fig. 7.2). Thus, the THz beam is focused into the sample at normal angle of incidence. The second arm (reference) contains a planar mirror, mounted on a translation stage for a delay of the image. The measured signal is the coherent superposition of the electric fields from these two arms. For a Gaussian beam, the complex electric field amplitude is given by:

$$E(r, z) = E_0 \frac{\omega_0}{\omega(z)} \exp\left(-\frac{r^2}{\omega^2(z)}\right), \ \exp\left(-ikz - ik\frac{r^2}{2R(z)} + i\zeta(z)\right) \qquad (7.1)$$

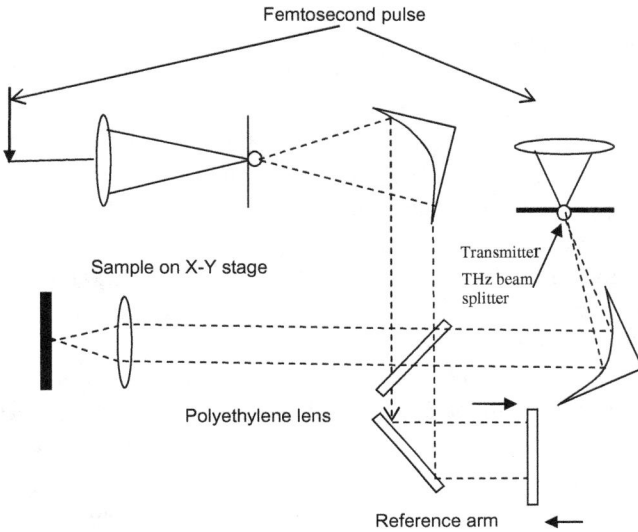

Fig. 7.2.   Imaging using interferometry.

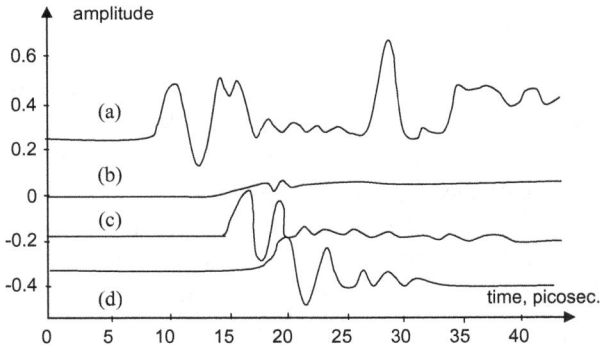

Fig. 7.3.   THz waveforms with (a) a delay between the signal and reference reflection; (b) the near cancellation of the two pulses; (c) the effect on the interferometric signal of the tape, and (d) the effect on the noninterferometric signal of the 45 μm-thick adhesive tape.

where $i\zeta(z)$ is the Gony phase (the longitudinal phase delay).

Due to the Gony phase shift acquired by the signal beam as it passes through a focus, it is 180° out of phase from the reference beam. If the sample reflects the THz pulse without distortion, the delay of the reference arm can be adjusted so that the two pulses destructively interfere at the detector. As a result, we detect almost no signal. Figure 7.3(a) depicts a waveform in which the sample and reference pulses are separated in time by a few picoseconds.

Figure 7.3(b) shows the wave front when the two pulses overlap in time, producing a new cancellation of the measured signal.

As an example, an adhesive tape attached to the cancelled surface is used to evaluate the effect of the tape being in the way of a THz pulse. The thickness of the artificial defect is approximately 45 μ or 1/7 of the coherence length of the pulse. The waveform in Fig. 7.3(c) shows the effect of the tape compared to the results in Fig. 7.3(b). Figure 7.3(d) shows the minute phase shift between the waveforms using conventional, noninterferometric imaging. The noninterferometric image is barely able to resolve the edges of the tape, while the interferometric image clearly shows the tape and air bubbles trapped under the tape. The destructive interference depends noticeably on the delay of the input signal as well as on the distortion that accompanies the signal.

The spectroscopic measurements are performed to identify the tablet. They require an analysis method of the THz waveforms. Usually, one measures the transmitted, time-domain waveform both with and without a sample present and then performs a Fourier deconvolution in order to extract the material parameters. The success of this method requires precise knowledge of the thickness of the sample, as both, the absorption and the phase delay vary exponentially with the sample thickness. Indeed, in many of these measurements, the dominant error in the optical constants arises from uncertainty in the thickness measurement. Also, analytical solutions do not exist to derive these constants from the measured electric fields, and we must employ numerical methods (Duvillaret *et al.*, 1996).

Very recently, Duvillaret *et al.* extended their work to also extract the thickness. However, only high index materials were considered (Duvillaret, 1999).

For the purpose of identification, the thickness of the password tablet's layers should be determined. The described identification process has used some of the results of the method proposed by Domey *et al.* (1999). The THz-TDS system provides a time-domain signal that contains not only the initial pulse transmitted through the material but also several subsequent pulses, resulting from internal reflections, which arrive at delayed times. The method to extract material parameters arises from the analysis of the multiple internal reflections described by the Fabry–Perot effect (Duvillaret *et al.*, 1996). A gradient search minimizes the difference between the model and the measured signal over a range of thicknesses (a model for penetration

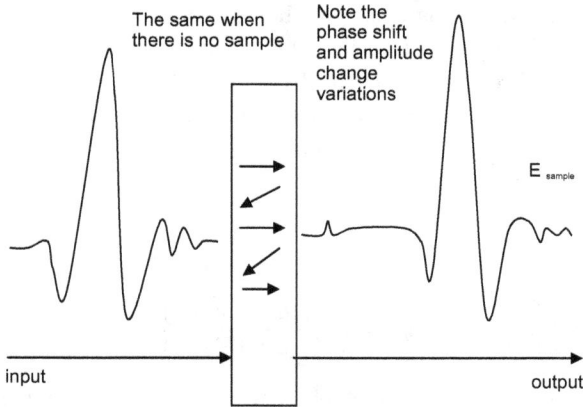

Fig. 7.4. A THz pulse transmitted through a homogeneous and planar material results in the initial pulse and several subsequent pulses due to multiple internal reflections.

(reflection)/refraction of the materials is built in advance) (Sokolnikov, 2006a). At each thickness that we assumed, we iteratively update the complex index of refraction function to minimize the total error. Once the complex index of refraction is identified for a particular thickness, a total variation metric measures the smoothness of the refractive index function. The total variation metric does not use any simplifying assumptions as in Duvillaret *et al.* (1996). The deepest local minimum of the total variation metric as a function of thickness identifies the predicted material thickness and the complex refractive index. For each pixel, the system receives a single waveform which is the result of the interactions between the THz pulse and the sample. One may also record a reference waveform without the sample *in situ*. Figure 7.4 shows a typical THz waveform without the sample $E_{\text{ref}}(t)$ characterized by a single-cycle pulse of approximately 1 ps in duration. Also shown, is a waveform measured with a planar sample in the THz path. This waveform $E_{\text{sample}}(t)$ (Fig. 7.4 on the right) shows a similar initial pulse due to the first transmission through the material. It also contains two smaller pulses caused by multiple internal reflections. We need to extract the difference in the pulse transmitted through the material. These differences, both in the time and frequency domains, form the basis for the information that we want to collect. The time-domain analog of the Fabry–Perot effect describes additional pulses. We see that all of the pulses in the waveform $E_{\text{sample}}(t)$ are time shifted, attenuated, and reshaped compared to the reference waveform $E_{\text{ref}}(t)$.

The Fresnel equations describe the transmission and reflection of the THz wave at each interface (Hecht, 2006). These are based on the material's complex index of refraction in the frequency domain: $\hat{n} = n(\omega) - jk(\omega)$, where $n(\omega)$ represents the real refractive index, $k(\omega)$ is proportional to the absorption coefficient, and ω is the angular frequency. The final equations at an interface between two layers are:

$$t_{ab}(\omega) = \frac{2\hat{n}_a(\omega)}{\hat{n}_a(\omega) + \hat{n}_b(\omega)}, \tag{7.2}$$

$$r_{ab}(\omega) = \frac{\hat{n}_b(\omega) - \hat{n}_a(\omega)}{\hat{n}_a(\omega) + \hat{n}_b(\omega)}, \tag{7.3}$$

where t_{ab} is the transmission coefficient of a wave at normal incidence (region a to b), and $r_{ab}(\omega)$ is the normal reflection in region a at a–b interface. As the wave moves through a material of thickness l, its propagation is governed by:

$$p_m(\omega, l) = \exp\left(\frac{-j\hat{n}_m(\omega)\omega l}{c}\right). \tag{7.4}$$

We neglect scattering (*e.g.* interface roughness) in our model and consider the THz path both with and without a sample in place. For the air, we have:

$$E_{\text{ref}}(\omega) = E_{\text{initial}}(\omega)p_{\text{air}}(\omega, x), \hat{n}_{\text{air}}(\omega) = 1.00027 - j0.3, \tag{7.5}$$

where x is the distance between the transmitter and receiver. It includes the small but measurable contribution of the refractive index from the air at standard pressure and room temperature (Sokolnikov, 2006b). We examine a planar, homogeneous material placed in the pathway of the THz radiation. Our iterative approach requires that the primary transmission and at least two multiples be present in the measured waveforms to solve for the free variables. The equation for the primary received signal and two multiples with the sample *in situ* are:

$$E_{\text{primary}}(\omega) = E_{\text{initial}}p_{\text{air}}(\omega, (x - l))t_{op}p_{\text{sample}}(\omega, l)t_{10}, \tag{7.6}$$

$$\begin{aligned} E_{\text{firstmultiple}}(\omega) &= E_{\text{initial}}(\omega)p_{\text{air}}(\omega, (x - l))t_{op}p_{\text{sample}}(\omega, l) \\ &\quad \times r_{10}p_{\text{sample}}(\omega, l)r_{10}p_{\text{sample}}(\omega, l)t_{10} \\ &= E_{\text{initial}}(\omega)p_{\text{air}}(\omega, (x - l)t_{01}r_{10}^4 p_{\text{sample}}^4(\omega, l), \end{aligned} \tag{7.7}$$

$$\begin{aligned} E_{\text{secondmultiple}}(\omega) &= E_{\text{initial}}(\omega)p_{\text{air}}(\omega, (x - l)) \\ &\quad \times t_{01}p_{\text{sample}}(\omega, l)t_{10}r_{10}^4 p_{\text{sample}}^4(\omega, l). \end{aligned} \tag{7.8}$$

To make our system of equations more tractable, we first add (7.6) through (7.8). The equations create a model of the measured waveform that contains all temporal signals:

$$E_{\text{compl}}(\omega) = E_{\text{ini}}(\omega)p_{\text{air}}(\omega, (x - l))t_{01}p_{\text{sample}}(\omega, l)t_{10}$$
$$\times \lfloor 1 + \Sigma_{k=1}^2 (r_{10}^2 p_{\text{sample}}^2(\omega, l))^k \rfloor, \tag{7.9}$$

We can now clearly see the multiples $FP(\omega)$ described by the Fabry–Perot effect. Dividing (7.9) by (7.5), we obtain:

$$\hat{H}(\omega) = \frac{E_{\text{compl}}(\omega)}{E_{\text{ref}}(\omega)}$$
$$= \frac{4\hat{n}_{\text{air}}(\omega)\hat{n}_{\text{sample}}(\omega)}{(\hat{n}_{\text{air}}(\omega) + \hat{n}_{\text{sample}}(\omega))^2} \left[\exp\left(\frac{-j(\hat{n}_{\text{sample}}(\omega) - \hat{n}_{\text{air}}(\omega))\omega l}{c} \right) \right]$$
$$* FP(\omega). \tag{7.10}$$

Equation (7.10) provides the transfer function for our model. The complex function $\hat{n}_{\text{sample}}(\omega)$ and l are the only variables. The deconvolution of our measured temporal signal (in angular frequency, ω):

$$H(\omega) = \frac{E_{\text{sample}}(\omega)}{E_{\text{ref}}(\omega)}. \tag{7.11}$$

Now, we want to compare the deconvolution of the measured signals $H(\omega)$ (Eq. (7.11)) with the model transfer function $\hat{H}(\omega)$ (Eq. (7.10)).

The real and imaginary parts of $H(\omega)$ are oscillating functions that produce many global minima for any error measure. Trying to avoid the above shortcoming, we use the magnitude and unwrapped phase information to provide a unique solution (Nuss and Orenstein, 1998). The algorithm should single out the phase for both the modeled and measured deconvolution similarity in order to make a valid comparison. In this process, we require that the phase must extrapolate to zero at zero frequency. The algorithm resets "0" Hz phase value to zero, and monotonically unfolds all the subsequent phases assuming that no two adjacent values differ by more than 2π.

The error is defined by taking the difference between the absolute values of the magnitude and unwrapped phase of the measured data and the model's:

$$mER(\omega) = |H(\omega)| - |H'|(\omega)|, \quad pER(\omega) = LH(\omega) - LH'(\omega). \tag{7.12}$$

The total error over all frequencies of interest is:

$$ER = \Sigma_\omega |mER(\omega)| + |pER(\omega)|. \tag{7.13}$$

The applied algorithm works in three steps. First, an initial guess of the thickness is made. Second, the initial functions for the complex index of refraction are calculated. Initially, a nondispersive material is assumed. Third, a gradient descent algorithm iterates the complex index of the refraction function in frequency until the total error no longer monotonically decreases. The algorithm records the final complex refractive index function, and repeats these steps for a range of different thicknesses. All the above steps are presented below in detail.

First, the algorithm bounds the upper and lower limits of the assumed thicknesses by:

$$l_{\text{upper}} = \frac{\Delta t_c}{n_1 - n_{\text{air}}}, \quad l_{\text{lower}} = \frac{\Delta t_c}{n_2 - n_{\text{air}}}, \quad (7.14)$$

where Δt is the time delay between the pulse in $E_{\text{ref}}(t)$ and the first pulse in $E_{\text{sample}}(t)$ of the measured signals. The parameters n_1 and n_2 limit the range of refractive indices considered with values of $n_1 = 1.2$ and $n_2 = 8$ covering almost all possible variations.

Second, the best initial starting function for the complex refractive index occurs when the first peak location of the temporal model's deconvolution is the same as the first peak's location of the measured deconvolution. The real refractive index and the thickness of a material control where the first peak exists in the temporal deconvolution. The imaginary index of refraction and the thickness affect the amplitude of the first peak. Assuming a nondispersive material (flat frequency response) for the estimated $n(\omega)$, the following equation governs the relationship between the thickness and real refractive index based on the location of the pulse in the measured signal's temporal deconvolution:

$$n = \left(\frac{\arg\max(|h(t)|)c}{l} \right) + 1.00027, \quad (7.15)$$

where l is the estimated thickness of the material selected in the first step and $\arg\max (|h(t)|)$ is the time index of the absolute maximum of the measured temporal deconvolution. This assures that the input signal is on at $t = 0$. The initial value for $k(\omega)$ is zero and it increases in increments until the absolute maximum of the modeled temporal deconvolution is less than or equal to the absolute maximum of the measured temporal deconvolution.

Third, after the initialization of $n(\omega)$ and $k(\omega)$, they are updated using a gradient descent algorithm (Haykin, 2001).

$$n_{\text{new}}(\omega) = n_{\text{old}}(\omega) + \varepsilon p E R(\omega); \quad k_{\text{new}}(\omega) = k_{\text{old}}(\omega) + \varepsilon m E R(\omega), \quad (7.16)$$

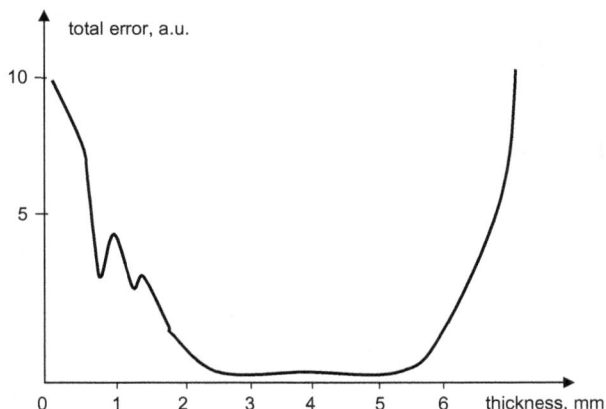

Fig. 7.5. Total error between the modal and measured signals for a 0.51 mm thick sample of silicon plotted over a range of thickness.

where ε is the update step size. It determines how much of an effect the magnitude and phase error have on the new values for the complex refractive index. (A reasonable value is $\varepsilon = 0.01$). The algorithm updates the complex index of refraction functions until the total error in (7.13) is no longer monotonically decreasing.

The implemented algorithm includes the three steps outlined above to generate a complex index of refraction function for a variety of thicknesses in question. Figure 7.5 shows the final error obtained from the described gradient descent algorithm at different thicknesses for a silicon sample. The shape of the plot is typical. The decaying exponential trend for small thicknesses is due to the exponential bias by (7.9) for the model. The growing exponential trend for large thicknesses is an artifact of the update size in the gradient descent algorithm. A smaller update coefficient (ε) would maintain the error curve near zero but dramatically increase the computational load. Under simulation, a global minimum is observed, however, experimentally obtained data, at best, provides the deepest local minimum. If the material under investigation has a relatively small real index of refraction, the concave error surface surrounding this minimum is extremely narrow. We need to determine which thickness and complex refractive index pair are the predicted properties of our sample. Instead of using the total error, we introduce a total variation of degree one (Odegard and Burrus, 1996).

$$D[m] = |n[m-1] - n[m]| + |k[m-1] - k[m]|, \quad TV = \Sigma D[m], \quad (7.17)$$

where the complex index of refraction is $\tilde{n}(\omega) = n(\omega) - jk(\omega)$ and the sum ranges between 250 GHZ and 2 THz.

The total variation (TV) measures the smoothness of the refractive index at the minimum identified by the gradient descent algorithm for each thickness. The deepest local minimum for TV is more easily determined than the deepest local minimum for the total error since the TV error surface provides a very broad concave region around the deepest local minimum. For most samples, we do not expect that the recorded complex index of refraction varies dramatically from one frequency sample to another, since the sampled frequency step size is relatively small. (Our temporal window width is approximately 50 ps with a sample rate of 5 fs which gives a frequency sampling of $\Delta f = 20$ GHz). Although the index may have strong variations with frequency, the majority of solid materials do not have special features compared to Δf (Nuss and Orenstein, 1998).

At each thickness, we use the recorded complex index of refraction at the final total error to calculate the total variation. As shown in Fig. 7.6, the complex index of refraction shows a marked reduction in oscillations at the proper thickness. The amount of rippling of $n(\omega)$ and $k(\omega)$ also decreases as l increases.

By identifying the thickness at which the deepest local minimum for total variations exists, the algorithm identifies the proper thickness. As we approach it, the oscillations in the complex index of refraction decrease. The general trend is a decrease in amplitude as the thickness increases.

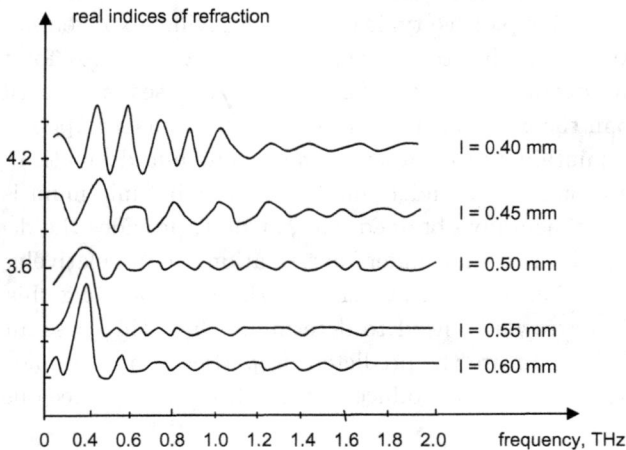

Fig. 7.6.   Final real indices of refraction obtained for different thicknesses.

This leads to the use of the deepest local minimum and the total variation of degree in (7.17) and (7.18).

Due to the large range of thicknesses to be considered, we apply the three steps outlined above to the three different thickness ranges and stepping distances. The first pass uses a coarse stepping distance over the full range of thicknesses identified in (7.13).

The next two passes use finer stepping distances over a limited range identified from the previous pass. The deepest local minimum of the total variation in each pass determines the center point for the next finer pass. In a limited number of situations, the final pass did not contain a minimum; therefore, we modified the total variation metric:

$$TV2 = \Sigma |D[m] - D[m = 1]|. \tag{7.18}$$

This modified total variation takes the absolute difference between adjacent points of (7.17). Using (7.18), we are able to amplify the variations in smoothness.

7.1.3. *Results for the example*

Several high index materials were examined and investigated for the general limits of the described approach, which requires that the primary transmission pulse and two multiples are identified in signal. Materials with low refractive indices may not have enough multiples above the signal-to-noise (SNR) limit. The results were produced for silicon wafers, GaAs and LiNbO$_3$ (ordinary axis).

Let us first consider the data for the silicon. Previous THz-TDS studies used silicon as a sample material since it has essentially zero absorption and a flat spectral response (*i.e.* $\tilde{n}(\omega) = 3.42 - j0$) (Grischowsky, 1990). Figure 7.5 shows that the total error plotted over the initial coarse sampling of thickness for a high resisting silicon wafer sample ($\rho > 10^4 \Omega.\text{cm}$).

Measurements made with calipers gave the thickness of approximately 0.51+/−0.01 mm. The deepest local minimum identified in Fig. 7.5 is approximately 30 μm off the measured value. The predicted thickness error is one tenth of the coherence length of the THz pulse.

Figure 7.7 contains the real and imaginary indices of refraction at the thickness 0.54 nm identified from the TV2 metric in (7.17). The solid lines indicate the predicted values. The line at 3.42 is the expected real refractive index (Grischowsky, 1990).

The predicted and expected values for the imaging components are superimposed. The real index is slightly low, since the predicted thickness

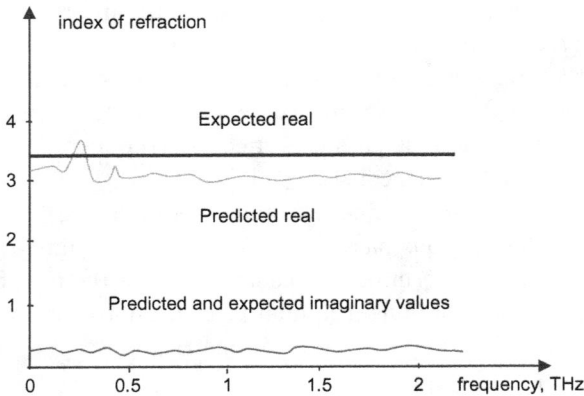

Fig. 7.7. Real and imaginary index of refraction for the predicted thickness identified in Fig. 7.5. Complex index of refraction for Si is $3.418 - j0$ (from a number of recent publications).

is rather high. The predicted real and imaginary values are independent of frequency as can be expected. The primary source of error is due to the alignment of the sample for which the THz beam should be at a normal.

The second sample, semi-insulating GaAs, has the following expected complex index of refraction (Domey *et al.*, 1999).

$$\tilde{n}(\omega) = \sqrt{11\left(1 + \frac{11.794}{8.055^2 - \upsilon^2 + j0.0719\upsilon}\right)}, \qquad (7.19)$$

where υ is frequency in THz. Due to the conventional manufacturing process, the complex index of refraction of GaAs varies noticeably from sample to sample, thus (7.19) is a good approximation.

The predicted thickness for GaAs was 0.38 mm compared to the measured thickness of 0.41 +/− 0.01 mm. Figure 7.7 shows the comparison between the predicted and measured refractive index values. Note that for GaAs variations of the index are more pronounced.

A small variation in the predicted thickness (Fig. 7.9) corresponds to a slight difference in the real index. Third, LiNbO$_3$ (ordinary axis) is examined. The results are given in Fig. 7.10. The dashed line represents the values for LiNbO$_3$ taken from literature. Figure 7.8 shows the variation of thickness for GaAs for the conventional manufacturing process.

A simulation is used to determine the limit of producing the parameters indicated above. In order to validate our approach at low values of the refractive index, a simulation system was created to yield the input and

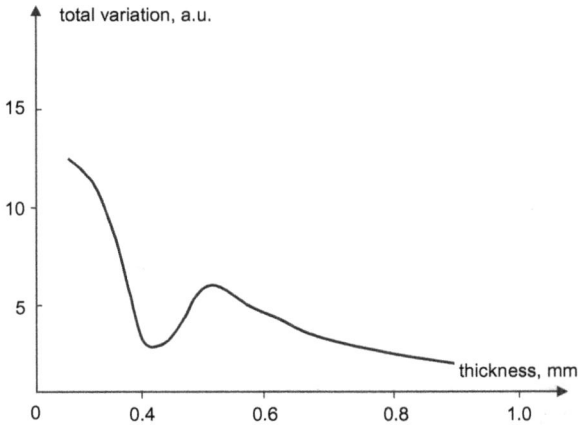

Fig. 7.8. Total variation (TV) measure for GaAs is shown with a 0.1 mm step.

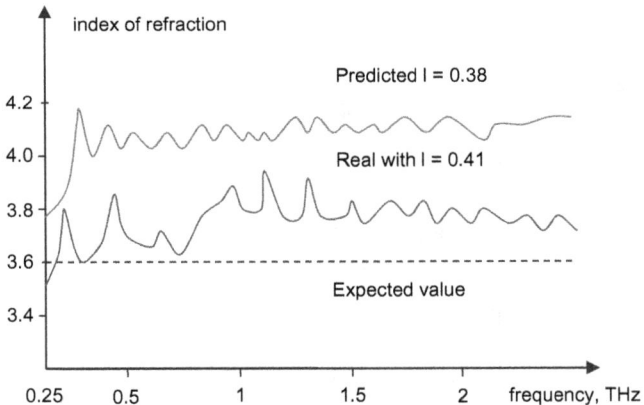

Fig. 7.9. Refractive index values for the predicted and measured thickness.

output time-domain waveforms. Rayleigh distribution models give a faster current rise and slower current fall in a photoconductive switch. The distribution was modified to smoothen the region around the origin to make it differentiable. The derivative of the modified Rayleigh distribution is used as a model for a single-cycle THz pulse.

From the SNR measurements of the THz system, we can expect a ratio of 1,500:1 using a relatively large number of average waveforms (>500).

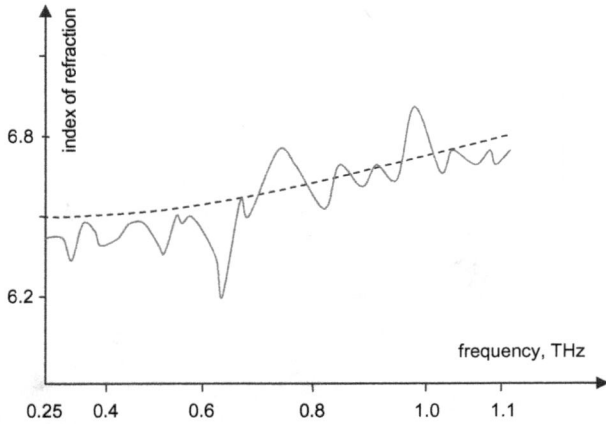

Fig. 7.10. Predicted real refractive index (smooth curve) and the real index generalized from literature for a sample of LiNbO$_3$ (ordinary axis) (Grischowsky *et al.*, 1990).

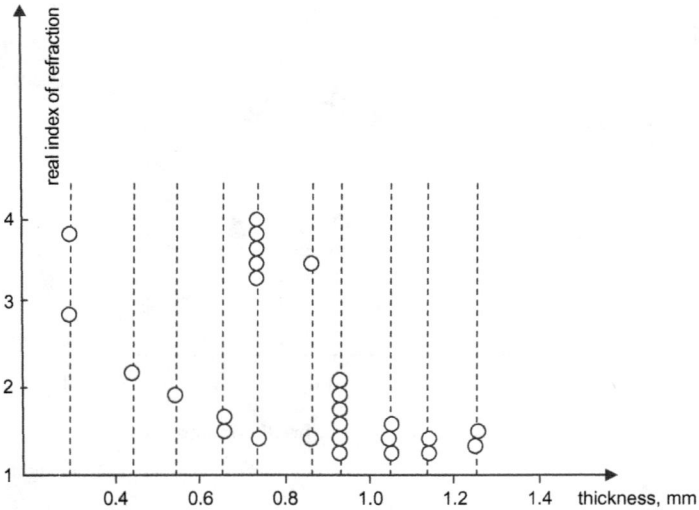

Fig. 7.11. Limits of the described method are shown for a simulated material at a signal-to-noise ratio of 1,500:1.

Using simulations and a captured noise signature, we determined the minimum real index determined by the algorithm. Figure 7.11 shows the range of parameters for which the method is applicable. The solid area represents the passing region. The area below is signal-to-noise limited.

The circles in Fig. 7.11 stand for the test cases that passed for each thickness given by the vertical dashed lines. At each point, the real refractive index was constant versus frequency while the imaginary component was zero.

7.2. Mobile THz systems

Portable (or mobile) THz systems are still at the stage of development although some prototypes may be considered almost commercial. Guo *et al.* at the Rensselaer Polytechnic Institute came up with "Portable THz ABCD", which is a THz radiation detection system that employs Air Breakdown Coherent Detection (THz ABCD) to detect THz radiation at up to 12 THz with 1000:1 signal-to-noise ratio. The system is powered by an external femtosecond (100 fs) laser and generates intense THz pulses by laser-induced air plasma. The prototype is enclosed in a 22 × 26 × 10 inch aluminum frame. THz pulses are generated by air plasma four-wave-mixing technique and detected by the ABCD, which is the inverse process of THz generation. Two delay stages are used in the system for common and fast scan modes. With nano-motion piezo motor, delay stage, the system gets up to one waveform per second — nearly real-time detection. The ABCD, however, is still in need of many technological improvements and developments.

Another example is "THz Spectroscopic Radar Mobile System for Detection of Concealed Explosives" sponsored by Small Business Innovative Research, Department of Defense, USA (SBIR). Physical Optics Corporation (POC) was contacted by SBIR DoD to design and manufacture a new 100 m stand-off THz Spectroscopic Radar (TSR), which uses THz molecular spectroscopy to detect the unique THz absorption wavelength signature of an explosive's out-gassed material. The design exploits molecular vibration modes in the THz region. To detect these specific explosives the TSR's design analyzes the retro-reflected THz signals from targets rather than the transmitted THz signals common to the conventional spectroscopy performed inside a laboratory. The TSR system combines a wide-band (1–20 THz) transmitter and a receiver spectrometer, built on a mobile station. Using a high-quality submillimeter wave radar to send THz probing signals with the rate of ~100 MHz, the system can reportedly identify not only the explosive but also the location of the target with spatial resolution better than 0.03 mm since the radar pulse width is faster than 100 fs. The average power of the source is higher (~100 W) than other presently known THz sources.

7.2.1. *Example: Mobile security surveillance system*

The following example is based on the THz Mobile Security System designed by the author. The design features a flexible electronic identification system that uses a mathematical algorithm that takes into account outside disturbances, which implies emulation of the encountered disturbances in order to curb them in subsequent measurements of the parameters of the identified object (Sokolnikov, 2006b).

Goals:

- To control certain areas from outside intrusion;
- To prevent an unauthorized exit or entrance (captives, prisoners, etc.);
- To control natural disasters' influence (floods, blizzards, etc.);
- To prevent wild animals from entering the controlled area;

What are the possible alternatives?

- Sentinel/guard — the most intellectual and versatile but costly and unreliable;
- Animals-guards — less intellectual, less costly, more reliable than humans;
- Microwaves — have been used ubiquitously; relatively high probability of identification error;
- Laser beams — costly, may be difficult to set up; controlled area is usually narrow;

Additional Means:

- Infrared (Night Vision);
- Fence, barbed wire, high voltage wire, etc.;
- Natural or man-made enclosures (a cave, buildings, etc.) — difficult to find and accommodate.

Possible Solutions:

- *Microwaves*: well-developed electronic implementation; wide area of detection but low resolution;
- *THz radiation*: wide possibilities of identification but difficulties with the electronic equipment; more complicated signal processing than (*e.g.*) for microwaves, more costly;
- *Laser Beams*: well-developed electronic implementation, narrow beam, narrow area of detection, costly.

Additional Issues:

- Sensitivity and resolution are not everything!
- Problem of discrimination between subjects/objects that cause similar disturbance of the probing signal;
- False alarms;
- THz images may not be diplomatic (the produced image may reveal details of a human body).

For this project, the frequency of the probe signal lies in the THz range. The portability was defined as the weight that an average person can carry (approximately, 17–20 kg). A set-up similar to the one shown in Fig. 7.2 was used with the resonance amplification power supply (Sokolnikov, 2007b) (see also Example: *Resonance Amplification* in Ch. 6) which satisfied the portability requirement. The emphasis was made on the reliability of object recognition. The following algorithm was suggested.

Time Series (Statistical Approach):

- No need for constant observations (as by usual video means);
- A model of the input signal is built based on the library of typical disturbances;
- The surveillance system is self-taught.

The algorithm for object identification is described in Fig. 7.12.

Historic background:

During the World War II, Brown and Holt designed a tracking system for fire control information to compute the location of submarines. The idea was

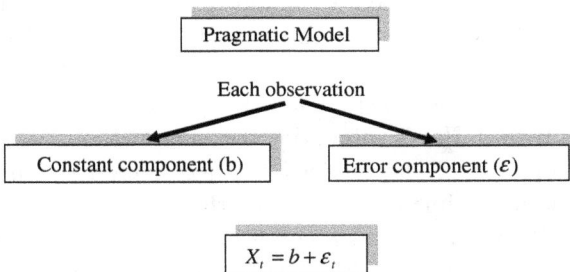

Pragmatic Model

Each observation

Constant component (b)   Error component (ε)

$$X_t = b + \varepsilon_t$$

Input signal representation in general:
b = stable in each segment of series (but may change slowly over time); ε = error;

Fig. 7.12.   Pragmatic model building.

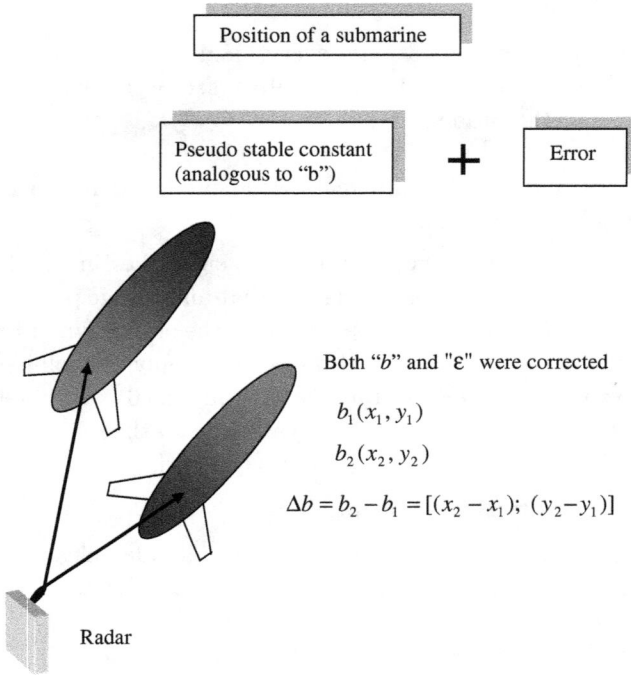

Position of a submarine

Pseudo stable constant (analogous to "b")   **+**   Error

Both "b" and "ε" were corrected

$$b_1(x_1, y_1)$$
$$b_2(x_2, y_2)$$
$$\Delta b = b_2 - b_1 = [(x_2 - x_1);\ (y_2 - y_1)]$$

Radar

Fig. 7.13.   Brown–Holt's submarine tracking.

to use well-defined location values and then use correction variables received from repeated measurements (repeated in time probe signals). The method allowed building pragmatic models. From the point of view of modern model building, Brown and Holt used a simplified but practical approach to an object tracking. Notwithstanding much more sophisticated mathematical apparatus (*e.g.* time series), the basic idea remains the same: there is a stable or pseudo-stable constant and a correction variable. Figure 7.13 illustrates the Brown–Holt's submarine tracking method.

The development of the model for object identification is described below. It is based on time series analysis that allows taking into account accidental factors such the weather condition (*e.g.* snow, hail, falling leaves, etc.). The model is corrected with each new weight assignment.

Weight Assignment I

- Isolate the true value of b;
- Compute a kind of moving average (the current and immediately preceding observations are assigned a greater weight than the next

observation);

$$S_t = \alpha^* X_t + (1 - \alpha)^* S_t - 1 \qquad (7.20)$$

If $\alpha^* = 1 \Rightarrow$ the previous observation is ignored.
If $\alpha^* = 0 \Rightarrow$ the current observation is ignored.

Weight Assignment II (Applying recursively to each success observation)

- Each new smoothed value (forecast) is computed as the weighted average of the current observation;
- The previous smoothed observation was computed from the previous value, etc.
- Thus, each smoothed value is the weighted average of the previous observation where weight-decrease exponent in-between produces intermediate results.

Developing a Model

The basic sinusoidal model is the foundation of the complex model:

$$Y_i = C + \alpha \sin(2\pi\omega T_i + \varphi) + E_i, \qquad (7.21)$$

where

$C =$ constant defining a mean level;
$\alpha =$ amplitude for the sin function; $\omega =$ frequency;
$T_i =$ a time variable; $\varphi =$ phase; $E_i =$ a model value for the dataset.

Conclusions for the example

- Flexibility of the algorithm allows the usage of different kinds of easily deployed systems;
- Principle of the weight assignment can be modified for particular condition of operation (such as different kinds of disturbance factors);
- Designed system is suitable for all electromagnetic ranges but it was primarily meant for THz;
- Resonance Amplifier allows the portability of the Mobile Surveillance System.

7.2.2. *Example: Compact THz system*

Recently a real-time THz imaging system, using the combination of a palm-size THz camera with a compact quantum cascade laser (QCL) was reported (Oda, 2012). The THz camera contains a 320×240 microbolometer focal

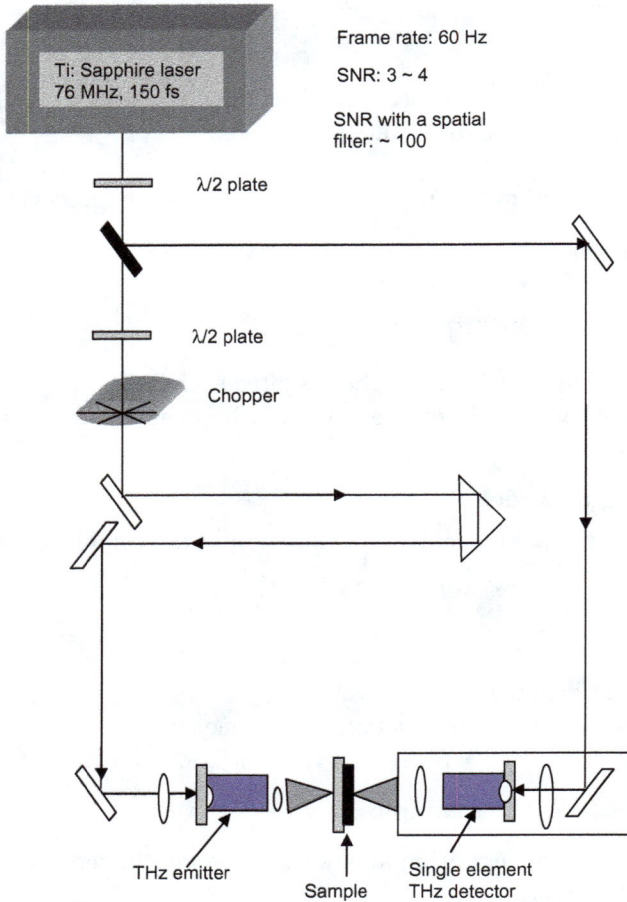

Fig. 7.14.   Mobile THz system's set-up.

plane array that works in the range of ~1.5–100 THz at 30 Hz frame rate. The cooled QCL has a range of ~1.5–5 THz, which determines the system's working range. The Japanese company, NEC Guidance and Electro-Optic Division has started initial production of a palm-size THz camera. The THz-QCL source operates in a pulse or CW mode. Below is given a schematic set-up of the described device.

The QCL active region was made into metal–metal waveguides (Fig. 7.15). The waveguide consists of a ground plane, on which the GaAs/AlGaAs active region was lithographically fabricated with a metal

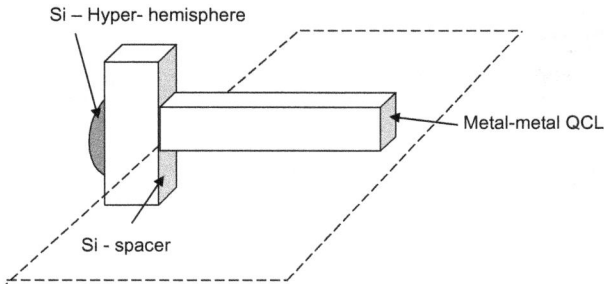

Fig. 7.15. Lens-coupled metal–metal waveguide.

contact on the waveguide's top. The dimensions of the waveguide were 1 mm × 100 μm × 10 μm. The waveguide design made possible almost unity optical mode overlapping with the gain medium, making possible a decrease in laser threshold currents and an increase of its operating temperature.

The QCL still needs to be cooled to 50 K. The cryogenic cooler is based on the Stirling cycle, which uses two linear motors driving two pistons. The pistons circulate a parcel of helium gas through a heat exchanger in a four-part pressure-volume cycle. Notwithstanding the necessity of cryogenic cooling, the imaging system is portable and is one of the first mobile THz identification devices.

7.3. THz identification of explosives

THz electromagnetic waves allow identifying explosive materials in a similar way the nonmetallic materials are identified (such as semiconductors and plastics). However, there are a number of aspects in which the explosives are different from the materials described in the previous section. In general, nonpolar, nonmetallic solids such as plastics and ceramics are at least partially transparent and reflective in 0.2–5 THz range. Nonpolar liquids are transparent as well, whereas polar liquids, such as water, are highly absorptive. Crystals formed from polar liquids are substantially more transparent because the dipolar rotations are frozen out. These crystals may exhibit phonon resonances in the THz range. Explosives are often not uniform-homogeneous substances, for example, humidity may influence their response to THz radiation. Experimental data suggest that many of the above materials (relevant to security application) have characteristic THz reflection and transmission spectra being a result of their crystal structure, the presence of impurities, etc. Table 7.1 contains an abbreviated summary of common explosive materials with their reported absorbance.

Table 7.1.   Collection of absorbance peak positions of some explosives (Federici *et al.*, 2005).

Material	Feature band center position frequency, THz
Explosive	
Semtex-H	0.72, 1.29, 1.73, 1.88, 2.15, 2.45, 2.57
PE4	0.72, 1.29, 1.73, 1.94, 2.21, 2.48, 2.69
RDX/C4	0.72, 1.26, 1.73
PETN[a]	1.73, 2.51
PETN[b]	2.01
HMX[a]	1.58, 1.91, 2.21, 2.57
HMX[b]	1.84
TNT[a]	1.44, 1.91
TNT	5.6, 8.2, 9.1, 9.9
NH_4NO_3	4.7

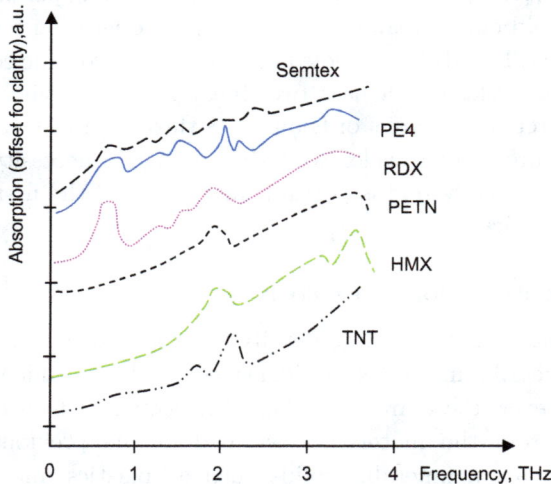

Fig. 7.16.   THz spectra of some common explosives (Federici *et al.*, 2005). Curves for PETN, HMX and TNT are given as an average for different sample configuration.

A dominant feature of the THz spectra is the sharp absorption peaks caused by phonon modes directly related to the crystalline structure. This result originates from the molecular vibrational modes and intramolecular vibrations associated, for example, with a common high explosive RDX (Huang *et al.*, 2003). Consequently, vibrational modes are unique and distinctive feature of the crystalline explosive materials. Note that there are several absorption features from 0 to 5 THz, which can be used to uniquely identify the explosives. Figure 7.16 shows a comparison of the THz spectra of a number of explosives (Federici, 2005). Note that the absorption features at 0.8, 1.5, 2.0 and others are present in Table 7.1 as well as in Fig. 7.16. It should be pointed out that the main absorption features of RDX and C4 are still present when RDX is mixed with a matrix material to make C4.

7.3.1. *Example: Detection of explosives*

The following examples consider the possibility of explosive detection concealed on personnel (Cook *et al.*, 2005). The THz explosive sensor utilizes a QCL. The THz explosives sensor is based on Differential Absorption LIDAR (DIAL). The principal of operation can be described as follows:

A given region is illuminated by THz radiation containing at least two distinct frequencies. This radiation transmits (or diffusively forward scatters) through the clothes, reflects partially off the skin or explosives (if present) and then scatters through the clothes a second time. The reflected power from each color is monitored by the sensor. The frequencies of the THz radiation are based on the THz spectra of the targeted explosives and are chosen so as to maximize the contrast in the presence and absence of the explosives. The ratio of the returned power from the two colors is measured by the sensor. A deviation from returned power ratios typical of human skin to returned power ratios typical of explosives will, therefore, determine the presence of concealed explosives. Thus, we can use sensors for nonimaging explosives identification.

7.3.1.1. *THz spectral signatures of high explosives*

In order to choose appropriate wavelengths for a practical THz explosives sensor, the influence of atmospheric air must be considered along with the spectra of the explosives. Figure 7.17 shows an experimentally measured spectrum of PETN (pentaerytritol tentranitrite) superimposed with an atmospheric absorption spectrum calculated by HITRAN. The same way, Fig. 7.18 shows an experimentally measured spectrum of HMX (cyclotetramethylene tetranitramine) along with the HITRAN calculated atmospheric absorption. From the above figures, we can obtain frequencies that fall within the windows between the peaks in the atmospheric absorption spectrum.

7.3.1.2. *Sensor set-up and operation*

The THz explosives sensor consists of three sub-systems: the THz transmitter, THz receiver, the computer and electronics. A set-up of the sensor components is provided in Fig. 7.19. The THz transmitter is based

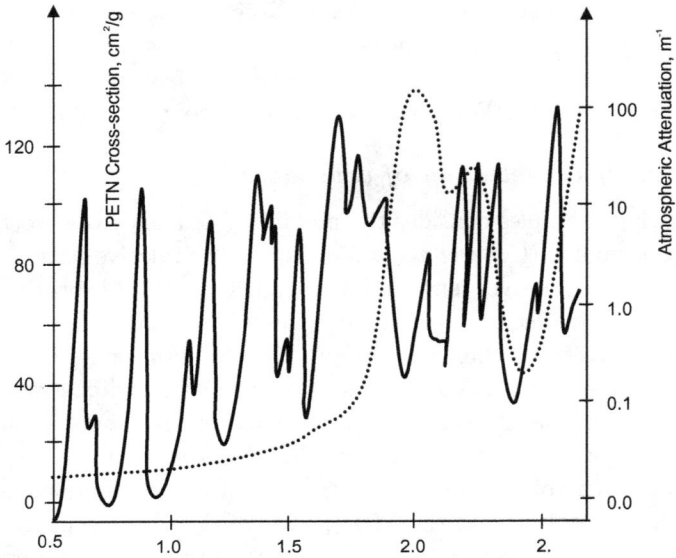

Fig. 7.17.   Experimental THz spectrum of PETN and a HITRAN modeled atmospheric absorption. The HITRAN simulations assume a temperature of 25°C and 30% relative humidity (Cook *et al.*, 2005).

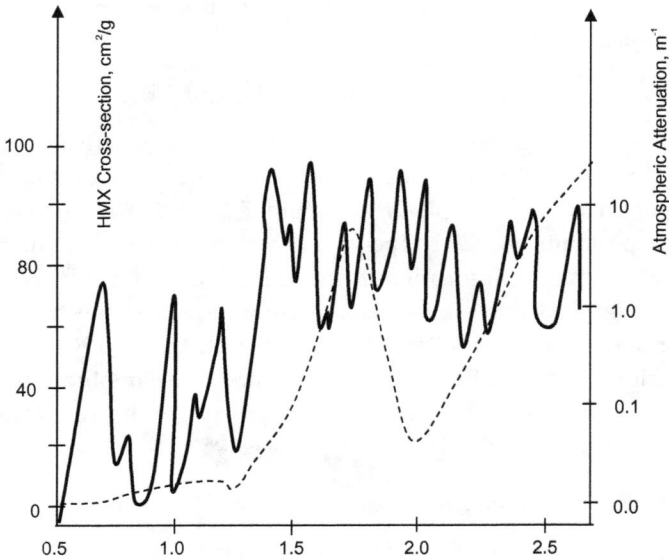

Fig. 7.18.   Experimental THz spectrum of HMX and a HITRAN modeled atmospheric absorption. The HITRAN simulations assume a temperature of 25°C and 30% relative humidity (Cook *et al.*, 2005).

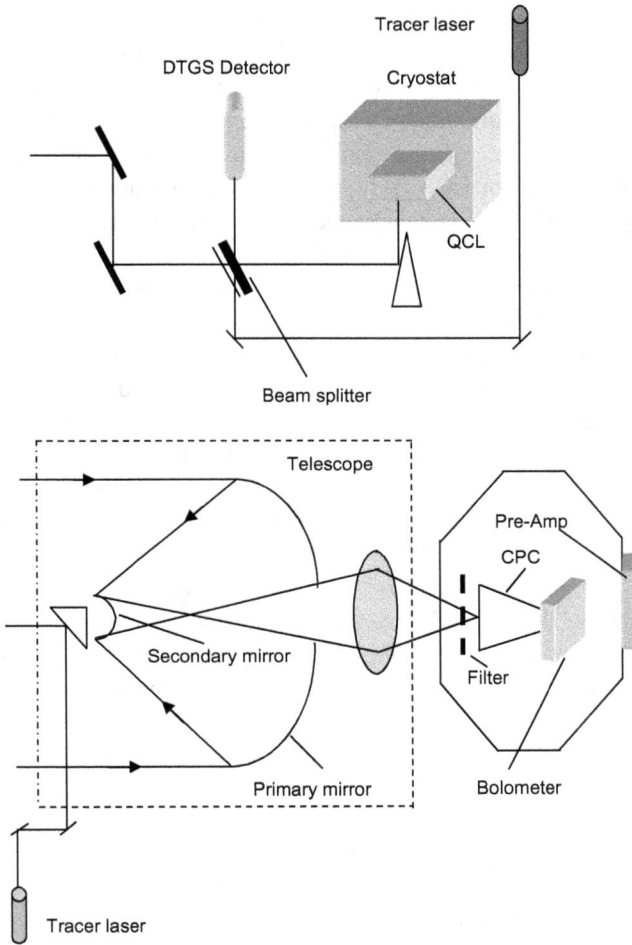

Fig. 7.19. (Part I) THz sensor explosives sensor set-up.

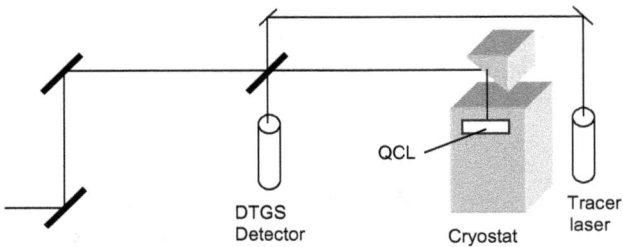

Fig. 7.19. (Part II) THz sensor explosives sensor set-up.

on a THz QCL. The described set-up consists of two QCLs provided by Tredicucci and co-workers, each mounted in liquid helium cooled in an open cycle cryostat. The radiation from each QCL is collimated by an off-axis paraboloid, and a beam splitter delivers a portion of the energy to a room temperature DTGS reference detector. A pair of steering mirrors directs the beam to the target. For each QCL, a visible range diode laser is coupled to the beam path. These tracer lasers are co-linear with the THz beams and are implemented to identify the target.

A parabolic 267 mm diameter primary mirror is used in conjunction with a hyperbolic 50 mm diameter mirror to form a Cassegrainian telescope with a 5 m standoff distance. A relay lens is then used to focus the collected radiation into a liquid helium-cooled bolometer (Infrared laboratories). A nonimaging Parabolic Concentrator (CPC) is incorporated into the bolometer. The CPC has an entrance aperture of 12.7 mm diameter and accepts rays that are f/3.8 or slower. The telescope provides the magnification of 4.9 and effectively collects radiation from a 62 mm diameter region. A visible light diode laser is aligned along the optical axis of the telescope. The region target by the sensor is, therefore, centered at the point where the three tracer lasers illumination converge.

Since the signal from the two laser sources must be detected by a single detector, each QCL is modulated at a distinct frequency. The signal from the bolometer is split into separate channels, which are demodulated at the certain frequency for each laser. The amplitude of the demodulated signals at frequencies 1 and 2 measured by the bolometer is denoted as B_1 and B_2 and the signals measured by the corresponding DTGS detectors are denoted by D_1 and D_2, respectively.

For each frequency, the normalized back-scattered signal is the ratio of the bolometer signal to the signal from the reference, *i.e.*

$$S_1 = \frac{D_1}{B_1}, \quad \text{and} \quad S_2 = \frac{D_2}{B_2}. \tag{7.22}$$

In the case of a two-color DIAL instrument, the ultimate observable is the following ratio of the normalized signals:

$$R = \frac{S_1}{S_2}. \tag{7.23}$$

The sensor will generate an alarm if R is outside an acceptable boundary.

An example of a practical application of a high explosive identification may be provided on the basis of the method described in Sec. 7.1 *THz*

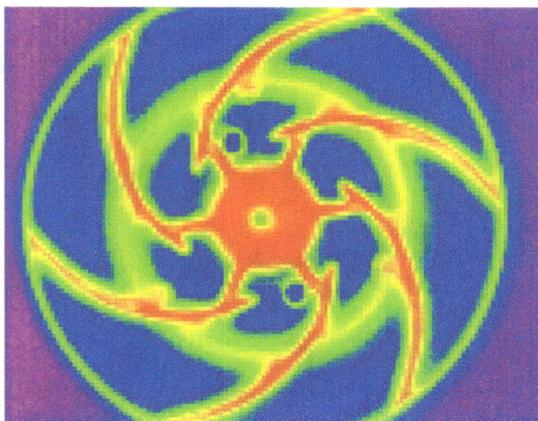

Fig. 7.20. THz image of the turbine-like structure with RDX explosive between the blades.

imaging of nonmetallic structures (applied by the author to the identification tablet). The measured index of diffraction of the target showed that the material inside the turbine-like structure (electrical rotor) was an explosive (Fig. 7.20).

The substance between the blades of the rotor (Fig. 7.20) is RDX — a high explosive (*high explosive* is an explosive material that is subject to detonation rather than *e.g.* deflagration, which is descriptive of *low explosives*). RDX was chosen because of its popularity with organized crime and terrorism. For example, as in many plastic explosives, the explosive in C4 is RDX. RDX is a colorless solid (often in a crystal-like form) of maximum theoretical density of $1.82 \, g/cm^3$. RDX is very stable at room temperature and can be safely tested in a laboratory. Sapphire was used as a substitute for RDX (with sapphire $n = 1.762 - 1.778$, which is close to that of RDX $n \sim 1.57$). The index of refraction for sapphire (RDX produced the same results) was determined using *Imaging Interferometry* set-up given in Fig. 7.2. The presence of the explosive substance was first determined as a discrepancy in the upper and lower limits of the thickness of the substance measurement between the blades in Fig. 7.20 applying Eq. (7.13). The real refractive index was the first peak in the temporal deconvolution. The n for RDX was found from Eq. (7.14). The total error for the thickness measurement of the sapphire (RDX yielded similar dependence) is given in Fig. 7.21. The graph is similar to the RDX shown in Fig. 7.16.

The real index of refraction of the explosive was found as the peak at about $f = 0.5 \, THz$ in Fig. 7.21.

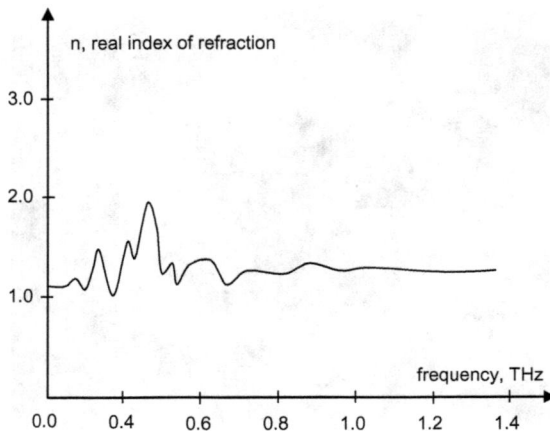

Fig. 7.21.   The real index of refraction for sapphire (similar for RDX) versus frequency of the source.

7.3.2. *Recent improvements of explosives identification*

So far the described methods of identification of explosives have often assumed to be the well-known THz-TDS technology that implies the analysis of the absorption or reflection spectra of the THz radiation, transmitted through or reflected from the substance. For which, the identification is based on the comparison of the absorption frequencies of the substance with those of the known explosives. However, this approach has limitations for reliable identification in a number of cases. For example, analogous spectra for some explosives and unrelated materials may be similar. A simultaneous measurement of all spectra of the substance can reduce the identification error. Such an attempt was made by V. Trofimov, *et al.* by using the reflected THz signal with measurement at different angles and different modes of operation (Trofimov *et al.*, 2012). In the transmission mode, the Spectral Dynamic Analysis (SDA) Method was applied for the identification of explosives with or without opaque covers (or wrappings), including mixtures of substances (explosive and nonexplosive materials) that have similar Fourier spectra in GHz and THz ranges. In particular, ceramic explosives (mixture of Al_3O_3 with other explosives, such as RDX and PETN) were investigated. The SDA discovered: (1) the decrease of the reflection at the frequency $\nu = 0.95$ THz during the main pulse. In the stand-off mode, ceramic RDX was detected under a polyethylene layer of up to 0.4 mm and under a cotton layer of about 0.3 mm. The weight of the pellet with pure or ceramic RDX was 800 mg. The absorption frequencies were $\nu = 0.8$ and 0.84 THz. The ceramic PETN was also identified under the same covering as above with the absorption frequency of $\nu = 1.5$ THz. So far, the above thicknesses of covering material represent the measure of penetration. It should be mentioned that the thickness of covers penetrated by the probing beam was sufficiently large for many identification cases. In other words, even now the depth of penetration with suitable results is suitable for many potential applications.

7.4. THz identification of concealed weapons

THz Identification of concealed weapons is supposed to improve the capabilities of conventional metal detectors and X-ray image devices. Steel-based weapons are easily detected by usual metal detectors, although the distance is usually less than 1 m. The metal detectors cannot identify the nature or design of the target. X-rays often require even shorter ranges to work properly and are harmful for the human being. In all above cases (but to a lesser degree distance-wise), THz radiation may be applied more successfully. The present challenges are the acquisition rate speed, resolution, size of the object/subject, the distance between the target and the detector and some others. At the present, a standoff distance of $\sim$2–3 m with the capability of visualizing some 0.5 m by 0.5 m area at an image rate of several frames per second are possible (Goyett *et al.*, 2008). Hidden metal objects can be detected under light clothing. One such set-up was demonstrated by T. M. Goyett *et al.* (2008). The authors primarily aimed to reduce long exposure time to produce adequate images and show a practical way to perform full body image scans at a near video rate. The source and receiver technology was of a standard type that included optically-pumped far-infrared lasers and room temperature Schottky diode receivers. The lasers and diodes operated at 1.56 THz. A frequency of 1.56 THz was chosen since it is located in one of the windows for transmission through the atmosphere where absorption of the power by water molecules is minimal. On the other hand, the chosen frequency should provide good resolution. In general, in order to receive video rate imaging at THz frequencies, the source frequency may be within the range of $\sim$0.3 THz and 1.6 THz (such as used for the above set-up by Submillimeter-wave Technology Laboratory (STL) at University of Massachusetts at Lowell). The current sizes of the molecular lasers components used by STL make them unsuitable for a portable THz imaging system. However, the availability of compact, coherent THz sources is critical for military and many special applications. In this respect, newer compact technologies such as QCLs operating at $\sim$1.0 THz and solid state multiplier sources with increasing frequencies ($>0.5\,$GHz) and diminishing weight and size are giving hope. It could be the next stage in development of the 1.56 THz set-up in consideration. One of the problems, however, solved by the set-up is the rate of imaging (2 frames per second) that makes it applicable to control admission in airports, custom houses, etc. The source set-up consisted of two 150 W, ultra-stable, grating-tunable CO_2 lasers that were used as the optical pumps for two far-infrared lasers. The CO_2 lasers were set to produce 9 μm and 10 μm wavelengths respectively. The outputs are then used to pump the laser transitions in the molecular gases difluoromethane and methanol at 1.5626 THz and 1.5645 THz respectively. The far-infrared lasers produce Gaussian modes with 10 mm FWHM and about 100 mW of power. One laser is used as the transmitter while the other serves as the receiver local oscillator (LO).

The optical design of the scanning system is shown in Fig. 7.22. It employs a LO, which is often used for nonlinear optics at optical frequencies of order of 10^{15} Hz. LO is an electronic device used to generate a signal usually for the purpose of converting signal to a different frequency using a mixer. This process of frequency conversion, also called heterodyning, produces the sum and difference frequency of the frequency of the LO and frequency of the input signal. LO acts as the power source to produce the optimal bias on the reference and receiver diodes. The LO signal passes through a series of Mylar beam splitters in order to illuminate both diodes simultaneously. The transmitter laser beam also passes through a series of Mylar beam splitters. A small amount of the transmitted power is combined with the LO signal on the reference diode.

The remaining transmitted power is diverted through the beam splitters and steered with a scan mirror for a raster scan of the subject. Unused power from both lasers is determined by THz absorbing material.

The theory of Gaussian beams has been used to calculate the exact positions for the optics in order to ensure that the field of view of the detector diode is overlapped with the transmitter beam. The system was designed to produce a two-way half power diameter of 7 mm on the subject at a standoff distance of 2.54 m.

Figure 7.23 shows the block diagram of the down-converting electronics that T. Goyett *et al.* used in their set-up. The heterodyne down-conversion employs the presence and interaction of a reference and received signal.

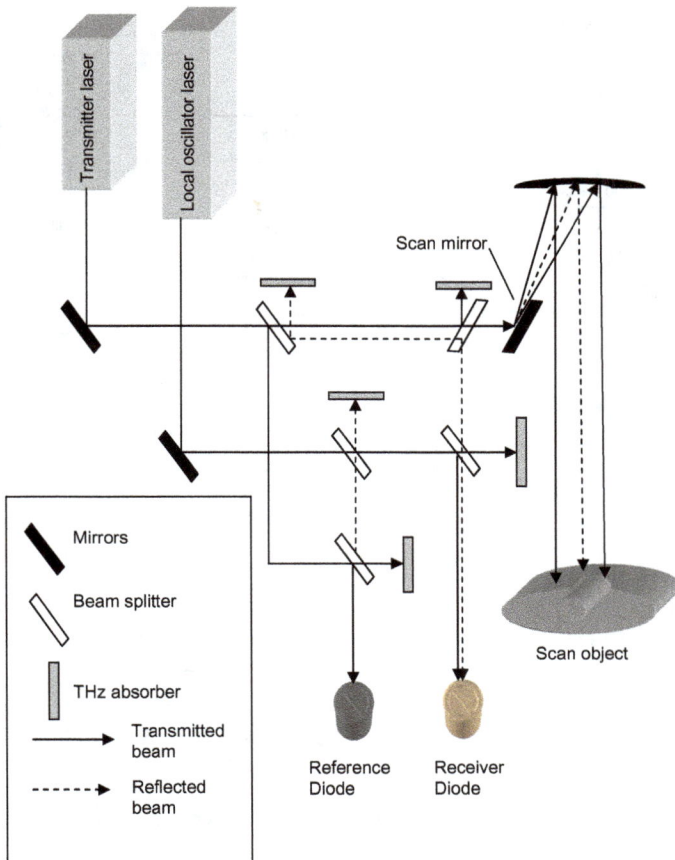

Fig. 7.22. Block diagram of the optical design for THz imaging system (concealed weapons identification).

Both signals are generated by 1.5 THz corner-cube-mounted Schottky diodes. The reference and receiver diodes in Fig. 7.23 are the same as in Fig. 7.22 but now they are shown with the signal processing electronics. The reference diode creates a beat frequency of 1.86 GHz between the transmitter and receiver laser. This signal is used as a reference to measure both the amplitude and phase of the received signal. The lock-in amplifier extracts the amplitude and phase information from the received signal. Since the difference frequency between the two lasers is 1.86 GHz, it is difficult to use a lock-in amplifier without further down-converting the

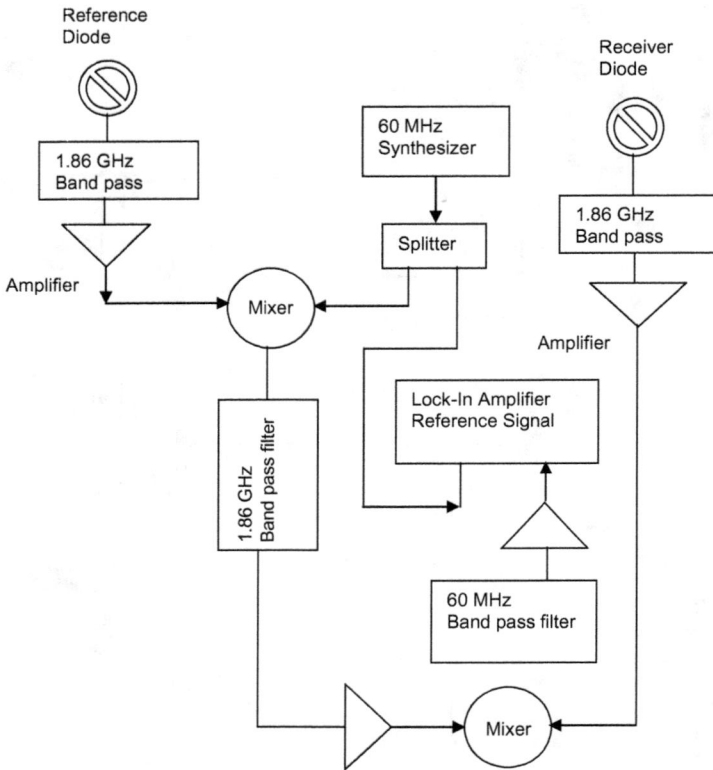

Fig. 7.23.   Block diagram of the down-converting electronics.

frequency. The reference signal is first shifted by 60 MHz to offset it from the received signal. The shifted reference signal is then mixed with the signal from the receiver in order to down-convert the signal to a stable 60 MHz frequency. The resulting signal is filtered and amplified before being sent into the lock-in amplifier. The 60 MHz synthesizer provides the reference frequency for the above amplifier. Thus, by this series of down-conversions the amplitude and phase information of the received signal is written into a very stable synthesized frequency that serves as a carrier.

7.4.1. *Image construction*

For image processing and data acquisition, the authors used National Instrument's Lab VIEW software. In order to provide a side-by-side comparison between the visible image and the THz imagery, the video camera was placed slightly off axis from the THz measurement path. A video frame was captured simultaneously with each frame of the THz scan. For the THz images typically a grid of 200 by 100 points requiring 20,000 complex data points per scan was required. This arrangement produced a data collection rate of 40 kHz per frame (2 frames per second). Complex amplitude and phase information was converted into the lock-in amplifier's output measured in volts with the time constant turned off, resulting in an effective data bandwidth of approx. 100 kHz. Effective scan areas of 0.51 m by 0.51 m could be achieved.

Once the THz scanner is set to work, the collected data is streamed to the computer disk. A video camera is mounted slightly above the large focusing mirror and monitored simultaneously. When each frame of the THz data is collected, the image from the video camera is stored along with the frame. As a result, both the video image of the target and the THz scan are correctly synchronized. The images provided by the authors were not high resolution only showing the contours of the detected pistol. However, it is important to point out that the target (pistol) is the only object to be seen — no anatomical details of the subject's body are visible. Thus, the identification can be specific, targeting only the points of interest.

Further, the imaging showed that the skin is highly reflective in the THz region. The features of the gun are much more efficient at back-scattering power than those of the hand that holds the weapon. Likewise, the material of the shirt (polypropylene) did not show very strong backscatter. However, it is likely due to the irregular surface features of the shirt. The gun's contours are visible at different angles, although the plastic trigger cannot be seen. Judging by the images, it is clear that the estimated amount of incident power ($0.5\,\mathrm{mW/cm^2}$) is sufficient to penetrate light clothing (cotton jacket, 0.7 mm thick) and scatter from the pistol. The jacket itself does not return more than a diffuse signal except for the zipper and clasp that yield a stronger reflectivity. Penetration of clothing was observed to be higher at 325 GHz, however, the resolution diminished at lower frequencies. Thus, an optimal balance between the penetration ability and resolution should be found. Figure 7.24 shows a simple scan of transmission and reflection through cotton fabric in the THz frequency region.

Note that the transmission through cotton has a significant variation from low frequency to high frequency. However, the reflection from cotton is relatively constant across the THz region, averaging less than 1%. The small reflectivity explains why the material shows only diffuse scattering. During the test scans, the power return observed on the spectrum analyzer was

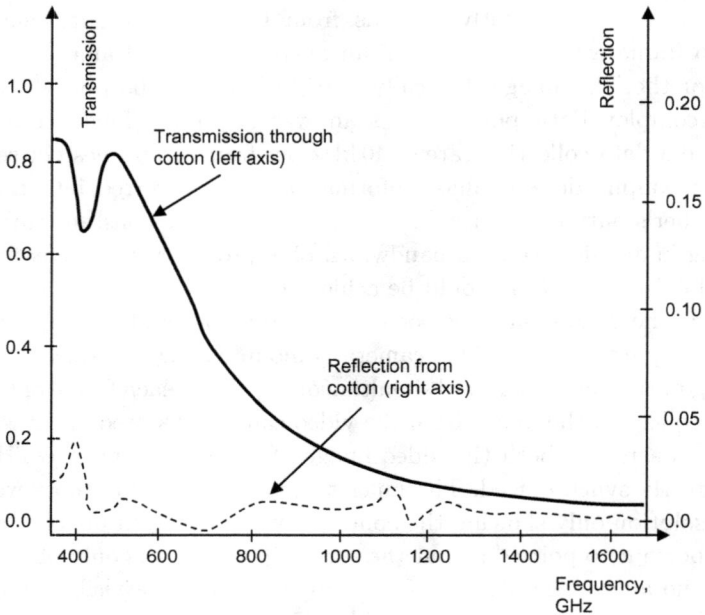

Fig. 7.24. Reflection and transmission dependence upon frequency.

60 dBm for the diffused scattering of the jacket. The reflectivity changes from material to material remaining approximately the above value for many materials/clothes made of fabric.

7.4.2. *Additional improvements for identification of metal objects*

The THz identification of concealed weapons poses a number of difficulties. One of them is the sensitivity of the detectors. There are several ways that may help to aid higher sensitivity and better identification. One is an external illumination source and another to employ more sensitive detecting surfaces of the bolometer (one of the most effective detectors) to stimulate higher absorption of the incoming THz radiation from the identified objects. Researchers from the Naval Postgraduate School in California have suggested a method of producing better images by means of improved detecting surfaces.

Most THz imaging systems are based on either antenna-coupled semiconductor detectors or cryogenically-cooled bolometers operating in

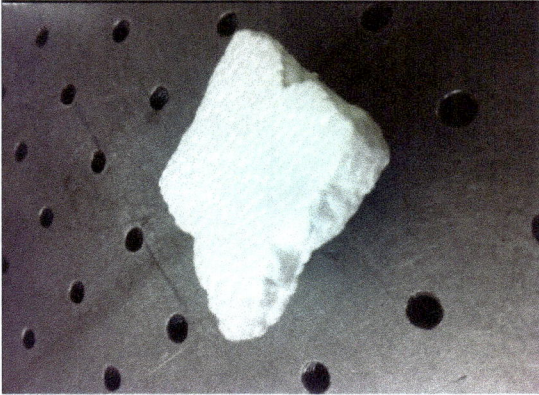

Fig. 7.25. (a) Styrofoam with a knife blade inside (courtesy of G. Karunasiri at Naval Postgraduate School, Monterey, CA).

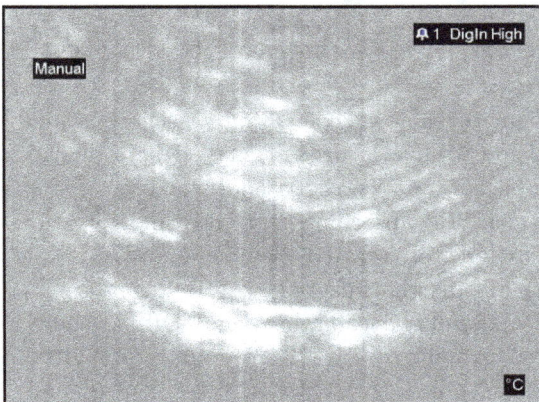

Fig. 7.25. (b) THz image taken by a microbolometer with THz optics (courtesy of G. Karunasiri at Naval Postgraduate School, Monterey, CA).

the scan mode. Since the 300 K (room temperature) background radiation does not contain appreciable power in THz frequencies for passive imaging, an external illumination source is typically needed for THz imaging. Figure 7.25 shows a THz image of a knife embedded in a piece of Styrofoam taken by a microbolometer camera with a Tsurupica-based lens in place of the standard Ge lens for better transmission of THz.

A THz QCL operating at 3.8 THz was used as the illuminating source for the images in Fig. 7.25. It can be seen that the THz radiation penetrates

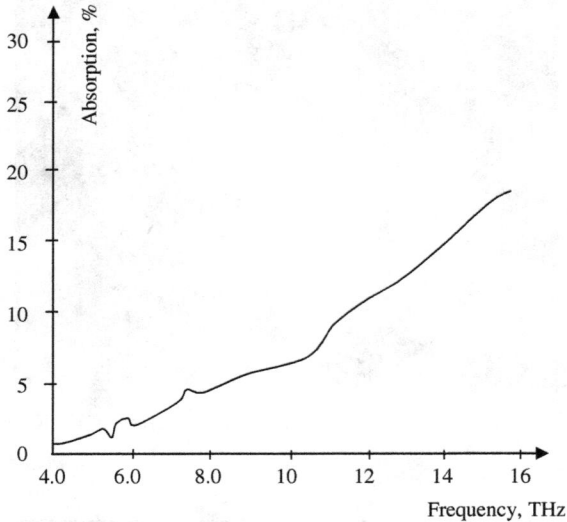

Fig. 7.26.   Measured absorption of a $1\,\mu$m of $Si_3\,N_4$ film.

through the foam revealing the location of the concealed weapon (knife blade). (The blade itself is a dark V-shaped figure surrounded by white segments). The sensitivity of the used bolometer was enhanced by the usage of enhanced THz absorbing materials for formation of pixels. The standard microbolometer infrared cameras employ a $0.5\,\mu$m-thick silicon nitride ($Si_3\,N_4$) film as the infrared absorbing material with about 80% absorption in 7–$13\,\mu$m wavelength range. Off-the-shelf microbolometer infrared cameras do not provide the optimum sensitivity necessary in our case.

Figure 7.26 shows the Si_3N_4 absorption dependence on frequency in the range that is typical for the present THz devices.

From Fig. 7.26, we can see that the absorption in the range of 1–$10\,$THz is less than 5% compared to almost 80% in the range of 7–$13\,\mu$m (which is long infrared).

Thin-metal films deposited on dielectrics provide good THz absorption due to resistive losses in the film. Hadley and Dennison predicted in 1947 that a free standing metal film in the air could absorb up to 50% of the incident infrared radiation. The absorption is given by (7.24):

$$A \approx \frac{2\sigma t}{2c\varepsilon_0 + \sigma t},\tag{7.24}$$

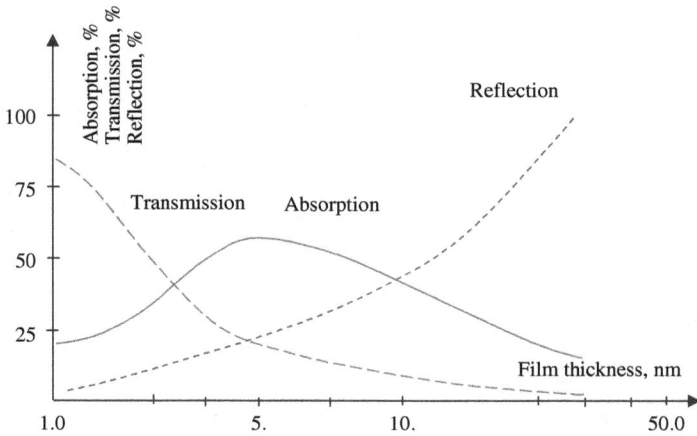

Fig. 7.27. Calculated transmission, reflection and absorption spectra as a function of film thickness (Alres *et al.*, 2011).

where σ is the film conductivity; t is film's thickness; c is the speed of light; ε_0 is the permittivity of space. Note that the absorption in (7.24) depends only on the product of film conductivity and its thickness. Also, since the conductivity of metal films is almost constant in the range of 1–10 THz (Smith *et al.*, 1998), the absorption is almost independent of THz frequency. The maximum absorption of 50% occurs at a film thickness of about 10 nm. Figure 7.27 shows the combined characteristics of the above thin films that include transmission, reflection and absorption.

A set of metal films, Cr and Ni were deposited on Si substrate. Both metal films showed an acceptable magnitude of absorption, which is presented in Fig. 7.28.

For applications involving external illumination using a narrow band THz source, the absorption can be further increased using metamaterial structures designed for resonant absorption of the narrow band THz radiation. In general, a metamaterials structure consists of a periodic array of metallic elements separated by a dielectric layer from a ground plane. As an example, a metamaterials structure comprises an array of Al squares separated an Al ground plane by a SiO_2 layer is shown in Fig. 7.29.

The achieved absorption coefficient versus frequency with the above metamaterials design is given in Fig. 7.30.

The thin metamaterials films that the researchers implemented for the detector were reportedly easily fabricated using standard MEMS and microelectronics materials such as SiO_2 and Al. Nearly 100% absorption

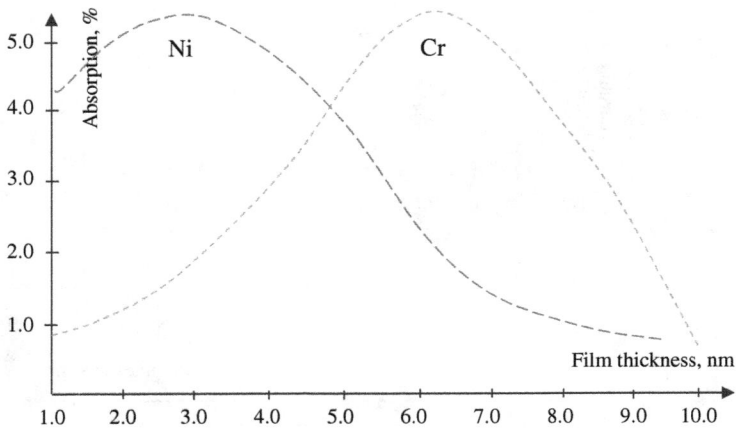

Fig. 7.28. Measured absorption for *Ni* and *Cr* films as a function of thickness (Alres *et al.*, 2011).

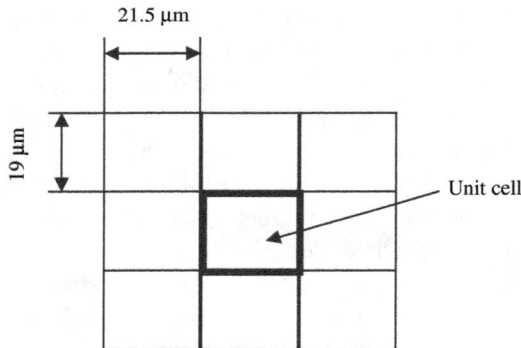

Fig. 7.29. SEM micrograph of the fabricated metamaterials structure on *Si* substrate (Alres *et al.*, 2011).

was achieved at the central frequency of 1–10 THz range. The sizes of the squares (Fig. 7.30) can be adjusted to the available QCL source. Such an adjustment may be performed in the range of 3–6 THz. The sizes of the squares may be adjusted to tune the peak absorption frequency. In this case, the smaller the square size, the higher the resonant frequency.

7.5. THz identification of illicit substances

While the literature on THz spectroscopy is quite extensive, the corresponding studies of THz sensing applications of illegal substances have

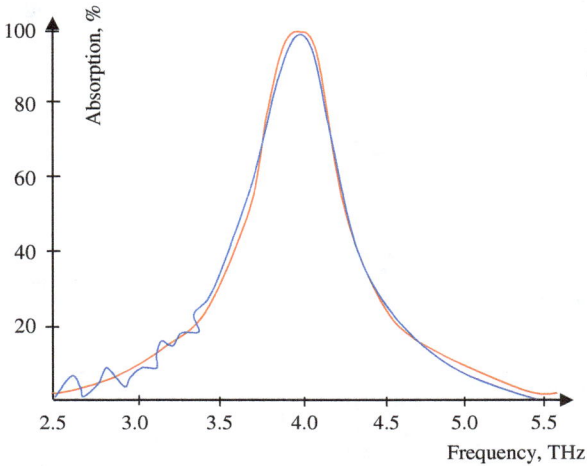

Fig. 7.30. Measured (red) and simulated (blue) THz absorption of the fabricated metamaterials thin-film structure (Alres *et al.*, 2011).

been much more limited. As with the detection of explosives, THz sensing shows great promise due to its ability to "see" through most packaging materials to probe the presence of drugs or drug-like substances that exhibit a characteristic THz spectral finger print. Competing detection technologies such as X-rays and millimeter wave imaging lack a spectral fingerprint in their respective electromagnetic frequency ranges to distinguish illicit drugs/narcotics from legal medication (Kawase, 2003). Infrared imaging has the capability to distinguish different drugs based on fingerprint spectra in the infrared region, but most packaging materials (*e.g.* envelopes, cardboard, etc.) are opaque to probing in the infrared. THz sensing allows identifying the characteristic features analogous to those for explosives (see Table 7.1). Kawase *et al.* came up with such results using a discretely tunable THz source (Kawase, 2003) to receive images of methamphetamine, MDMA (DL-3.4-methyllenedioxymetamphetamins) and common aspirin as a reference. The major THz absorption peaks in the 1.3–2 THz range in these drugs are shown in Table 7.2 (Federici *et al.*, 2005).

About 20 mg of each drug was individually placed in polyethylene bags and hidden in an envelope. THz images were recorded at seven different THz frequencies and analyzed using component spatial pattern analysis (Watanabe *et al.*, 2003). The analysis showed an error of roughly 10%. Kawase *et al.* have also demonstrated that they can identify (extract) the spatial patterns of the different drugs even when they were mixed

Table 7.2. The major THz absorption peaks for some common drugs (1.3–2 THz range).

Material	Feature band center position frequency, THz
Drugs:	
Methamphetamine	1.20, 1.70–1.80
MDMA	1.40, 1.80
Lactose α-monohydrate	0.54, 1.20, 1.38, 1.82, 2.24, 2.57, 2.84, 3.44
Icing sugar	1.44, 1.61, 1.82, 2.24, 2.57, 2.84, 3.44
Co-codamol	1.85, 2.09, 2.93
Aspirin, soluble	1.38, 3.26
Aspirin, caplets	1.40, 2.24
Acetaminophen	6.50
Terfenadine	3.20

together or stacked on top of each other. (This is an important consideration for security applications where people would intentionally attempt to confuse a THz sensing system). Kawase *et al.* also reported characteristic THz spectra for D-ampheamine, L-ephedrine and L-methylephedrine and over-the-counter drugs: L-methamphetamine, acetaminophen (Tylenol) and caffeine (Kawase *et al.*, 2003). Thus, a wide spectrum of drugs is available for identification even at the present moment.

One of the methods showing promise as imaging modality for stand-off detection of drugs is interferometric imaging (Stand-off identification appears to be more practical for security applications since for monitoring large numbers of people at airports, etc. a distance of least several meters is required.). The interferometric THz imaging design does not require necessarily a particular coherent or incoherent source of THz waves. It is flexible enough to utilize an electronic THz source, a laser-based illuminating source or incoherent ambient THz radiation. A long-term advantage is that interferometric imaging may produce more information than a single line-of-sight system. With repeated measurements, it is possible to apply imaging processing and computational techniques to multiply images and THz sources to aid noise and false alarm reduction. The interferometric approach, for example, was successfully implemented by the author to nonmetallic structures and explosives (see earlier in the chapter, Sec. 7.1.2 and *Example* "Detection of explosives").

7.6. Prospects and conclusions

As it was mentioned earlier, the security and military applications often need easily transportable identification equipment. For example, in order to implement a field (portable) version of the considered explosive sensor, improvements in both the THz laser sources and detectors are required. In particular, the requirements for liquid helium cooling, and consumable cryogens in general, must be eliminated. This would require more mature QCLs with improved temperature performance. Even though considerable progress has been achieved for THz QCLs at frequencies near 5 THz, liquid nitrogen cooling is still required. Furthermore, the development of QCLs at frequencies below 3 THz (for wavelength longer than 100 μm) is necessary both to target the previously reported resonances in the high explosives and to provide effective transmission through the atmosphere to penetrate clothes material. The production of the above QCLs poses a substantial challenge even when liquid helium cooling is used.

Methods using Imaging Interferometry (given at the beginning of this chapter for measuring parameters of nonmetalic structures) can be successfully implemented (as in the example above). However, one of the challenges with explosive identification is the explosive nonuniformity. For example, dynamite is a mixture of highly sensitive nitroglycerin with sawdust, powered silica, or most commonly diatomaceous earth that act as stabilizers. Thus, the refractive index, for example, is not the same throughout the target area. Also, error in n determination remains substantial due to frequency variation.

While crystalline, high energy explosives have characteristic THz fingerprint spectra that can be used to identify these threats, home-made ammonium nitrate bombs and other improvised explosives could pose a challenge to THz security applications, since these materials can have featureless THz spectra below 3 THz.

Lastly, one of the biggest challenges to THz security applications is stand-off detection. As the stand-off distances increases, one must consider the effect of the humid atmosphere, dust, smoke, etc. as well as possible barrier materials. As stand-off distances, picosecond pulsed measurements become problematic. In order to overcome the attenuation losses of barrier materials and the atmosphere, higher power sources need to be developed. In conjunction, compatible low-noise THz receivers need further development. Concealed weapon identification is one of the THz applications that proved to be successful at usual standoff distances (several meters and farther). The current signal processing and optical equipment allow receiving real-time images that are target-selective (*e.g.* without anatomical details of the subject-carrier of the weapon). However, the challenge is still the resolution, penetration ability, the size of the scan, the availability of compact THz sources as well as some other aspects. Nevertheless, even the current level of the technological development provides practical set-ups for the weapon identification. The provided descriptions outline all the important features of the successful sensor design. Practical stand-off distances can be evaluated and the effects of different types of clothing (or other materials) on the ability to detect explosives can be directly measured.

So far, the advances of technology have determined new THz applications. More recently, though, the applications themselves have impacted upon technology development. For many THz applications to become everyday reality, new compact sources and detectors are required, as well as compact power supplies. This means that further advances in electronics and nanotechnology should be made.

References

Alres, F, B Keamey, D Grbovic, NV Lavrik and G Karunasiri (2011). Strong terahertz absorption using thin metamaterials structures. *Applied Physics Letters*, 100.

Cook, DJ, MG Allen, BK Decker, RT Wainner, JM Hensley and HS Kindle (2005). Detection of high explosives with THz radiation. *30th International Conference on Infrared and Millimeter Waves*, Williamsburg, VA.

Domey, T, J Johnson, D Mittleman and R Baraniuk (1999). Imaging with terahertz pulses. *Applied Optics*, 38.

Duvillaret, L, F Garet and J Coutaz (1996). A reliable method for extraction of material parameters in terahertz time-domain spectroscopy. *IEEE Journal of Selected Topics in Quantum Electronics*, 2, 739–746.

Duvillaret, L, F Garet and J Coutaz (1999). Highly precise determination of both optical constants and sample thickness in terahertz time-domain spectroscopy. *Applied Optics*, 38, 409–415.

Federici, JF, B Schulkin, F Huang, D Gary, R Barat, F Oliveira and D Zimdars (2005). THz imaging and sensing for security applications — explosives, weapons and drugs. *Semiconductor Science and Technology*, 20.

Goyett, TM, JC Dickinson, KJ Linden, WR Neal, CS Jpseph, WJ Gorreatt, J Waldman, R Giles and WE Nixon (2008). 1.56 Terahertz 2-frames per second standoff imaging. *Proc. SPIE*, 6893.

Grischowsky, D, S Keiding, M VanExter and C Fattinger (1990). Far-infrared time-domain spectroscopy with terahertz beams of dielectrics and semiconductors. *Journal of the Optical Society of America B*, 7, 2006–2015.

Haykin, S (2001). *Adaptive Filter Theory*. Englewood Cliffs, NJ: Prentice Hall.

Hecht, E (2006b). *Optics*, 2nd edn. Reading, MA: Addison-Wesley.

Huang, F, B Schulkin, H Altan, J Federici, D Gary, R Barat, D Zimdars, M Chen and D Tanner (2003). Terahertz study of 1,3.5-trinitro-s-triazine (RDX) by time-domain spectroscopy and FTIR. *Applied Physics Letters*, 83.

Kawase, K, Y Ogawa and Y Watanabe (2003). Non-destructive terahertz imaging of illicit drugs using spectral fingerprints. *Optics Express*, 11.

Nuss, M and J Orenstein (1998). Terahertz time-domain spectroscopy (THz-TDS). In *Millimeter and Submillimeter Wave Spectroscopy of Solids*, G Gruner (ed.). Heidelberg, Germany: Springer-Verlag.

Odegard, JE and CS Burrus (1996). Discrete finite variation: A new measure of smoothness for the design of wavelet basis. *Proc. of ICA SSP*, 1467–1470.

Oda, N, AWM Lee, T Ishi, I Hosako and Q Hu (2012). Proposal for real-time THz imaging system, with palm-size THz camera and compact quantum cascade laser. *Proc. SPIE*, 8363.

Smith, DY, E Shile and M Inokuti (1998). *Handbook of Optical Constants of Solids*, ED Palik (ed.). Academic.

Sokolnikov, A (2006a). Adaptive non-intrusive terahertz identification. *Proc. SPIE*, 6212.

Sokolnikov, A (2006). Mobile security surveillance system. *Proc. SPIE*, 6201.

Sokolnikov, A (2005). Resonance amplification of the probing signals in optical coherence tomography (OCT). *Proc. SPIE*, 5881.

Sokolnikov, A (2007a). *Remote Identification of Foreign Subjects*. Singapore: World Scientific.

Sokolnikov, A (2007b). THz identification of humans and concealed weapons for law enforcement, government, and commercial applications. *Proc. SPIE*, 6538.

Trofimov, V, S Varentsova, M Szustakowski and N Palka (2012). Efficiency of the detection and identification of ceramic explosive using the reflected THz signal. *Proc. SPIE*, 8363.

Watanabe, Y, K Kawase and T Ikari (2003). Component spatial pattern analysis of chemicals using terahertz spectral imaging. *Applied Physics Letters*, 83.

Chapter 8

Medical and Other Applications
of THz Radiation

Biomedical sensing is one of the most rapidly developing of all terahertz (THz) applications. Many biomedical developments are suitable for security and defense applications. One of the examples is DNA identification used, in particular, in forensic medicine. THz spectroscopy (initially one of the first THz applications) allows determining the development of basic THz fingerprints of simple molecules, such as water, carbon monoxide and ozone, where THz waves can probe a range of energy transitions that correspond to the excitations of rotational, vibrational and translational modes in complex organic molecules, including biomolecules. Using THz absorption spectroscopy, the structure and potentially the dynamics of complex molecules such as proteins, DNA, RNA, and larger biostructures such as cell clusters and bacterial spores can be identified and analyzed. This application of THz radiation opens possibilities for fast DNA analysis in both areas of disease detection and forensics. Biosurveillance and chemo-surveillance are becoming increasingly important to protect against biological and chemical attacks.

Another useful aspect is that breathing modes of DNA caused by stretching of hydrogen bonds between the two DNA strands — vibrations, twisting and global stretching modes — are excited by radiation in the 0.1–10 THz band. This range is available for THz spectroscopy which provides information about DNA structure and dynamics. In DNA analysis, THz absorption spectroscopy can be used, for instance, to distinguish A and B configurations. Additionally, THz spectra differ for DNA from different organisms for various single mutations.

Recent advancement in the fields of ultra-fast femtosecond pulsed lasers and electro-optics have culminated in practical sources of coherent broad-band THz pulses and facilitated room-temperature detection that has enabled THz imaging for medical applications as well as for a number of other fields. The dominant method of imaging is THz-pulsed imaging (TPI), where the laser pulse is used for the generation and coherent detection of THz. In this method, the THz pulse is scanned across the sample, a pixel at a time, in a reflection or transmission mode by parabolic mirrors (Fig. 7.2).

The TPI is an advantageous method for medical imaging since there is no ionization hazard for biological tissues and the Rayleigh scattering of electromagnetic radiation. The intensity is many orders of magnitude less for THz wavelength than for the neighboring infrared and optical regions of the spectrum. The THz frequencies correspond to energy levels of molecular rotations and vibrations of DNA and proteins, and these may provide characteristic fingerprints to differentiate biological tissues in a region of the spectrum sensitive to water and exhibit absorption peaks due to stretching modes at 6 THz and librational ("librational" implies a molecule's ability to rotate freely as opposed to the motions under the forces of the neighboring molecules) modes at 19.5 THz. This makes the technique very sensitive to the degree of hydration, which can indicate tissue condition.

In TPI, a series of broadband pulses reflected from or transmitted through structures within a sample are measured, providing a large amount of temporal and spectral information for each pixel. Fourier transform methods have been used to access the spectral information from the time-domain signal; however, these methods are not capable of determining the temporal characteristics of frequency components. Alternatively, wavelet methods (wavelet analysis was briefly introduced in Ch. 3), which are designed for analyzing pulses, are ideal for the broadband THz pulsed radiation and provide additional information about the temporal occurrence of spectral content.

Wavelet is a wave-like oscillation with amplitude that increases from zero, increases, and then decreases back to zero. It usually may be called a "brief oscillation". Generally, wavelets are purposefully designed to have specific properties that make them useful for signal processing. Wavelet can be combined, using a "shift, multiply and sum" technique called *convolution*, with portions of a known signal to extract information from the unknown signal.

For example, a wavelet can be created that has a set frequency and duration. If this wavelet were to be convolved at periodic intervals with a signal that has that frequency/duration pattern, then the results of these convolutions would be useful for determining the presence in the signal of our preset wavelet pattern. Mathematically, the wavelet will resonate if the unknown signal contains information of similar frequency. This resonance phenomenon is the central concept for many practical applications of wavelets.

As a mathematical tool, wavelets can be used to extract information from many different kinds of data, including images. In general, to provide a comprehensive analysis of the data, we need sets of wavelets. In addition, a set of "complimentary" wavelets can deconstruct the data without gaps or overlaps so that the deconstruction process is mathematically reversible. Thus, sets of complimentary wavelets are useful in wavelet-based compression/decompression algorithms where it is desirable to recover the original information with minimal loss.

More specifically, a wavelet is a mathematical function used to divide a given function or continuous-time signal into different scale components. A frequency range may be assigned to each scale component. Each scale component can then be analyzed with a resolution that matches its scale. A *wavelet transform* is the representation of a function by wavelets. The wavelets are scaled and translated copies (known as "daughter wavelets") of a finite-length or fast-decaying oscillating waveform (known as the "mother wavelet"). Wavelet transforms have advantages over traditional Fourier transforms for representing functions that have discontinuities and sharp peaks, and for accurately deconstructing and reconstructing finite, nonperiodic and/or nonstationary signals.

The key advantage that the wavelets have over time domain and Fourier-based techniques is provided by reflection geometry. In the Fourier analysis, the frequency content of the waveform as a whole is considered, whereas in the wavelet analysis the frequency content is considered at a specific time. Thus, the waveforms that consist of multiple reflected pulses analyzed easily using wavelets, since each reflected pulse has a localized response in the wavelet domain.

Summarizing, a wavelet represents a square-integrable function with respect to either a complete, orthonormal set of basis functions, or an over-complete set or Frame of a vector space (also known as a Riesz basis), for the Hilbert space of square integrable functions. Further, wavelet transforms are classified into discrete wavelet transforms (DWTs) and continuous wavelet transforms (CWTs). Note that both DWT and CWT are continuous-time (analog) transforms. They can be used to represent continuous-time (analog) signals. The difference between DWT and CWT is that CWTs operate over every possible scale and translation whereas DWTs use a specific subset of scale and translation values or representation grid. In general, wavelets have two areas of application within TPI: for signal pre-processing, that is, compression and noise reduction, and data analysis. In this chapter, we shall only consider the analysis.

For the purpose of identification, wavelet methods in data analysis involve transforming the reference and sample waveforms into the wavelet domain, and then correlating the results to obtain the closest match for translation, which is the temporal parameter, and scale, which is inversely proportional to frequency. The reference pulse itself maybe used as the mother function (or prototype wavelet) with cross-ambiguity function.

Further discussion seeks to validate the use of wavelets in TPI by comparison of optical properties (refractive index and absorption coefficient) derived for a known test object by wavelet methods and Fourier transform techniques. The data were collected in transmission rather than reflection to enable a direct comparison between the Fourier and wavelet methods.

8.1. Data analysis — Example: Wavelet analysis

A step-wedge Duraform polyamide sample (with selective laser-sintering process) was used for measurements. The prototype had six thicknesses with steps: 0.3 mm, 1 mm, 2 mm, 4 mm, 6 mm, and 7 mm. Figure 8.1(a) shows an example of the transmitted pulses through various thicknesses of a step wedge. Figure 8.1(b) shows the spectral content of the transmitted pulses.

A THz waveform of the transmitted pulses was obtained for each pixel of the step wedge image. Pulse time-delay and attenuation estimations (relative to the reference pulse) were made using Fourier-based and two wavelet-based techniques, namely a Wide-Band Cross Ambiguity Function, and a Wavelet-Based Wide Band Cross Ambiguity Function (WB-WB CAF).

An area of interest of around 80 pixels was defined for each step, and the results within each area were averaged to produce a simple estimation for the time delay and attenuation of that step. From these, the refractive coefficients were calculated in the following sections of this chapter. Since

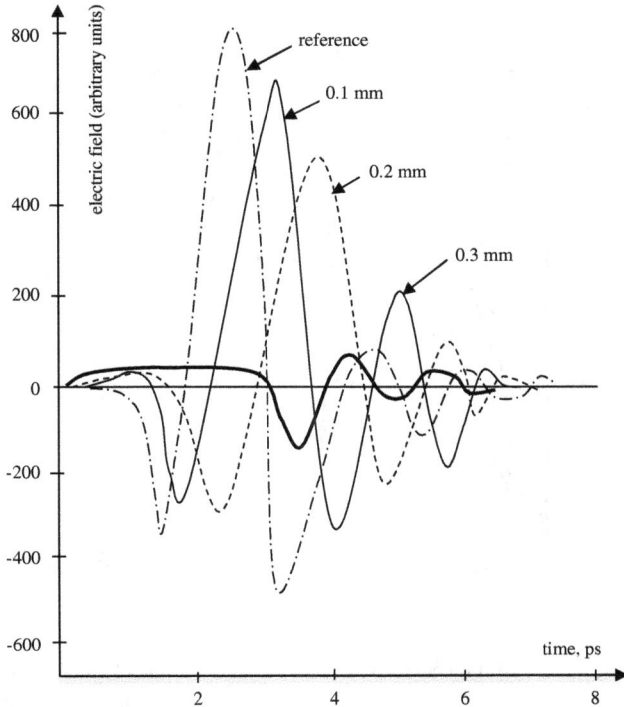

Fig. 8.1. (a) Example of broadband THz pulses delayed and attenuated by different thickness of the sample.

the absorption coefficient is frequency dependent, the relative transmission was estimated at a number of frequencies (using Fourier techniques) and scales (using the wavelet techniques).

8.1.1. *Calculation of refractive index (for the Example)*

The estimated time delay was plotted against the known sample thickness, and the line of best fit calculated by linear regression. The relationship between thickness and time delay is

$$TD = \frac{(n-1)x}{c}, \qquad (8.1)$$

where TD is the time delay of the pulse, n is the refractive index, x is the thickness and c is the speed of light in a vacuum. The gradient, m, of the best fit corresponds to $\frac{(n-1)}{c}$, and by rearrangement $n = 1 + mc$.

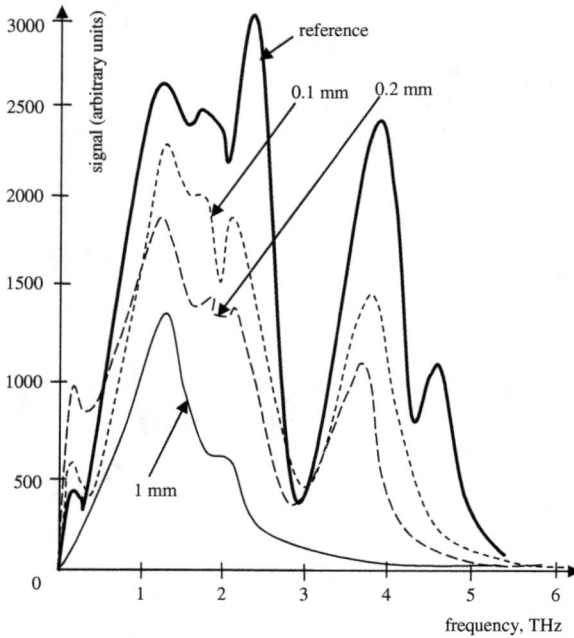

Fig. 8.1.   (b) The corresponding spectral content of each pulse.

8.1.2. *Calculation of absorption coefficient (for the Example)*

The absorption coefficient was calculated from the plot of the natural logarithm of the relative transmission against the known sample thickness. The reflection losses (using Fresnel coefficients) and the refractive index (from 8.1.1) were also taken into account. The absorption coefficient, α, can then be derived directly form the gradient of the best fit regression line, based on the *Beer–Lambert law*:

$$\ln\left(\frac{I}{I_0}\right) = -\alpha x, \tag{8.2}$$

where I_0 is the incident intensity, I is the transmitted intensity, and x is the sample thickness. It was assumed that the measure of attenuation based on the correlation value from the cross-ambiguity function was directly proportional to the relative transmission.

8.1.3. *Results for the Example*

Values for the refractive index of the step wedge using time-domain analysis, WBCAF, and WB-WBCAF are shown in Table 8.1.

Figure 8.2(a) shows a plot of step-thickness versus estimated time delay for the three techniques. Figure 8.2(b) shows the Fourier-calculated absorption coefficient versus frequency. For the WBCAF and WB-WBCAF, Fig. 8.2(c) shows inverse scale (proportional to frequency) against an estimate of absorption coefficient based on the attenuation. This parameter has been normalized for comparison with the Fourier-based results by dividing through by the median value.

The initial results for wavelet techniques are encouraging in the above example compared with a traditional time-series analysis of the data, although time-series analysis is more universal and applicable in the general case. Both cross-ambiguity functions give comparable time-delay estimations, and hence similar refractive index estimations. Wavelets are

Table 8.1. Calculated refractive index and absorption coefficient versus analysis method.

	Time domain	WBCAF	WB-WBCAF
Refractive index	1.603+/−0.004	1.60+/−0.001	1.59+/−0.01

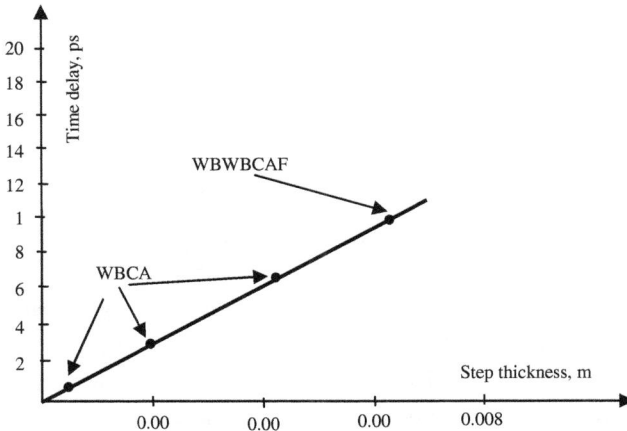

Fig. 8.2.   (a) Comparison of time-delay estimation methods.

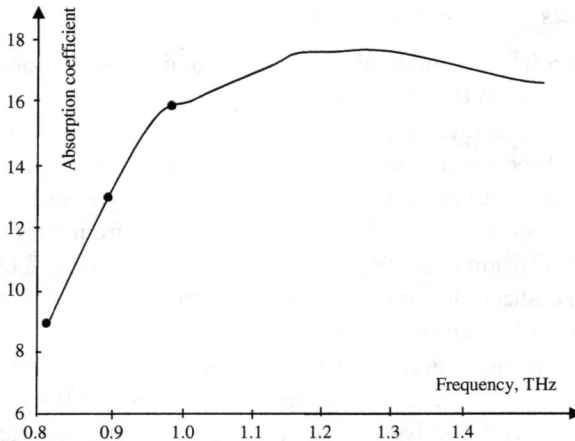

Fig. 8.2.   (b) FT calculation of absorption coefficient for Nylon wedge.

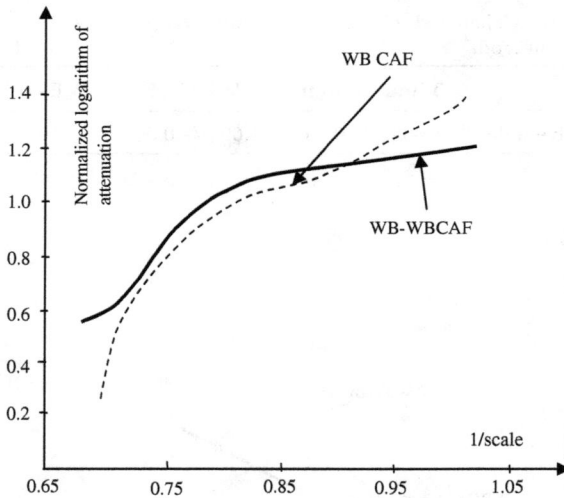

Fig. 8.2.   (c) Wavelet absorption coefficient estimation versus inverse scale.

already used in the field of ultrasound for time-delay estimation, and so we would expect them to perform well in a number of applications with THz pulses. Furthermore, the cross-ambiguity functions indicate a similar profile of the absorption parameter against frequency. The scale value of the wavelet transform corresponds to a frequency band, and as such it is not possible to convert the scale value directly to a single frequency, which precludes direct comparison based on frequency. With the range of scales analyzed, the results are promising, and the range of scales can be expanded to cover the full spectral content of the pulses.

Summarizing, wavelet methods may be useful for analyzing THz reflection data, such as the signals recorded from interfaces between the components of skin, the epidermis, dermis and hypodermis (or subcutaneous tissues), which can be individually differentiated by their signature. This is important because the ability to differentiate between the layers may be helpful in identification in forensics in case one or two layers deteriorate and become not suitable for analysis.

Promising dental applications with THz imaging also arise from the above-mentioned reasons and are helpful for forensics as well. Further, medical applications are considered more in detail.

8.2. Dermatology

The skin constitutes one of the largest organs of the body and is accessible for reflection imaging. It is composed of several layers, principally the epidermis, dermis and hypodermis (or subcutaneous tissues). This layered structure is attractive for reflection imaging as there are interfaces between materials with different physical properties that cause reflection of THz pulses. Measurements on model human dermis have shown that the skin constituents could be differentiated by TPI. Characteristic changes are seen in images of the skin using ultrasound or Magnetic Resonance Imaging (MRI) in the presence of skin diseases such as psoriasis where there is thickening of the epidermis (El Gammal *et al.*, 1999). The potential offered by TPI as a measurement technique is that its practical implementations may be cheaper and more compact than MRI and may not require contact with the skin as does ultrasound.

Surface wounds may arise from burns or trauma, whilst chronic, nonhealing wounds can result from venous ulcers, pressure sores and diabetic ulcers. Wound healing has three distinct phases. The first stage is hemostasis and includes the inward movement of inflammatory cells. The second phase is the formation of granulation tissue and the third is the remodeling of that granulation tissue causing contraction of the wound. In a criminal investigation, it may be important to know not only the final stage but the stages previous to it. THz sensing is one of the several methods that allow determining the condition of all the layers of skin.

Surface optical imaging techniques are increasingly used to differentiate between benign and malignant lesions from their characteristic morphology. However, the depth of penetration is also important, because, for example, cancers that have not yet penetrated the dermis will not have had the opportunity to metastasize to other areas of the body. Reflection TPI is expected to provide a measure of depth assuming knowledge of the lesion absorption properties. However, this is an application for which there are several competing technologies. One of them, *Optical coherence tomography (OCT)* uses light in the visible and infrared regions and produces cross-sectional images with a resolution of about 10 μm for a penetration depth of 2–3 mm. The images are analogous to B-mode ultrasound images. The applications for OCT are very similar to those for TPI, but TPI is expected to perform better because of reduced scattering and a comparable resolution. Recently, a number of innovations have been implemented that improved the resolution of OCT. One of them is the usage of resonance amplification of the probe signal (Sokolnikov, 2005). Thus, OCT remains a strong contender of TPI's.

Figure 8.3 depicts the cross-sections of a capillary taken with or without resonance electronic amplification. Figure 8.3(a) gives only a small portion of the arterial cross-section and Fig. 8.3(b) the entire section is presented.

Microscopic techniques include epiluminescence microscopy and confocal microscopy. Confocal microscopy allows cross-sectional imaging of planes parallel to the skin surface, *in vivo* (Webb, 1996) and three-dimensional reconstruction can be used to generate sections in the more familiar vertical direction used in histological slides.

Near-infrared wavelengths are preferred to optical wavelengths because of the lower scattering. They can achieve lateral and axial resolutions of 0.5 μm and 2 μm respectively within a small sampling volume, but only to a depth of 0.5 mm.

In contrast, experimentalists have shown that TPI penetrates to at least 1.5 mm in a model dermis system suspended in bovine calf serum (Bezant, 2000) and is associated with even less scattering (Han *et al.*, 2000).

Nonoptical techniques under investigation for skin imaging include MRI and ultrasonography. Ultrasonography has proved valuable in assessing skin tumors by their composition and extent, and early work suggests that THz imaging has similar capabilities (Woodward *et al.*, 2001). The potential of TPI, however, to provide both a structural image and spectroscopic information means that it may be able to complement the alternative techniques. For example, ultrasound methods cannot distinguish

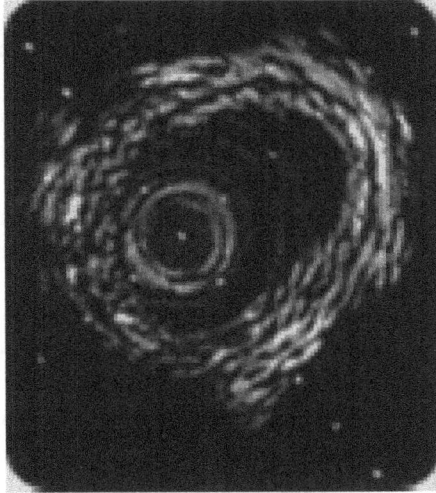

(a) **Without electronic amplification.**

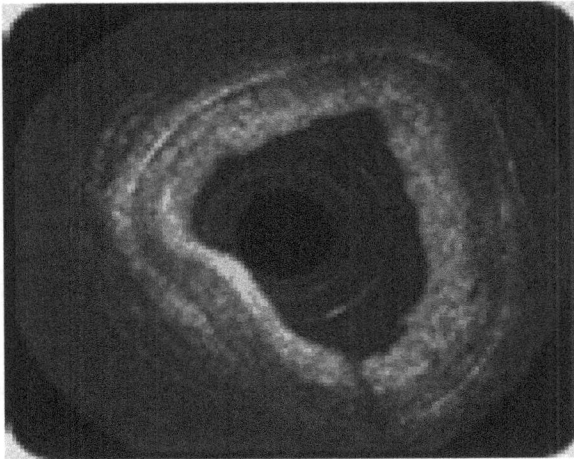

(b) **With electronic amplification.**

Fig. 8.3. Comparison between the resolution capabilities with or without electronic amplification.

between benign and malignant lesions because naevi can be very thin and not identifiable with ultrasound probes. Also, it is sometimes hard to differentiate the tumor itself from inflammatory changes induced by tumor. It has been shown (Woodward *et al.*, 2001) that TPI can be used to distinguish between normal skin and basal cell carcinomas *in vitro*. Initial work on an optical method involving a model of color formation within the skin has shown promise for demonstrating skin architecture (Cotton *et al.*, 1997). These techniques may also be valuable at THz frequencies, alone or in combination with the optical methods. The water content of the stratum corneum can affect the physical characteristics of the skin, including its barrier function, mechanical properties and fluid penetration. Alternative methods used for measuring hydration were surveyed by Berardesea and Borroni (1995) and this is a very active area of research. The main imaging contender is MRI. However, MRI is expensive and special coils are needed to achieve the spatial resolution required for imaging skin, so there is a role for TPI as an alternative method. It is known that Fourier transform infra-red spectroscopy (FTIR) provides valuable spectroscopic information (Stuart, 1997) and reduced scattering at the THz frequencies may enhance these results.

8.3. Hard tissues diagnostics (dentistry, surgery, etc.)

Recent developments in THz technology allow THz analysis of hard biological tissues (primarily bones and teeth). High-resolution images are important for forensics and surgical applications for in-field and hospital treatment of casualties. One of the methods developed recently is Scanning Near-field Infrared Microscopy (SNIM) (Schade *et al.*, 2005) mentioned earlier in the book. SNIM together with synchrotron radiation source (SRN) at BESSY (Schade *et al.*, 2005) (was tested on a number of biological tissues, in particular on teeth in sub-THz range. With well-established X-ray techniques, buried lesions in bone tissue (*e.g.* in teeth) can be imaged by a contrast change. However, the density contrast is weak in comparison to SNIM and buried structural defects (or ailments), such as lesions, are not visible until much later in the deterioration process. Specific tests on demineralization identification in teeth were done using the THz SNIM together with the coherent synchrotron radiation source at BESSY. If caries lesions are detected early enough, they can be arrested without the need of surgical intervention.

In Fig. 8.4, a tooth slab contains a lesion visible from the surface. This tooth specimen was used to receive confocal and near-field images of the specimen given in Fig. 8.5.

With the SNIM method, buried caries lesions can be imaged by a contrast change due to demineralization in the particular tooth region

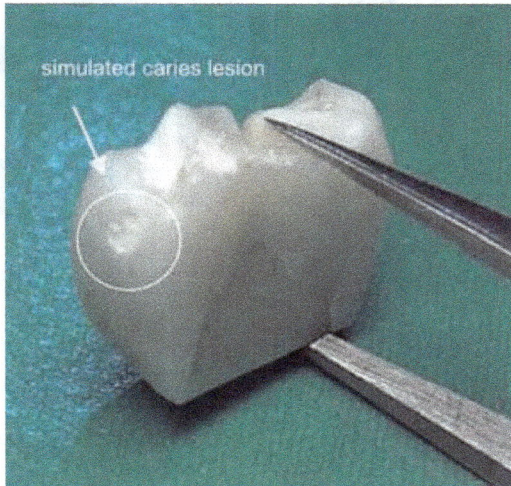

Fig. 8.4.   Photograph of a 2.7 mm thick human tooth slab. The buried lesion Is indicated by an arrow on the left side of the slab.

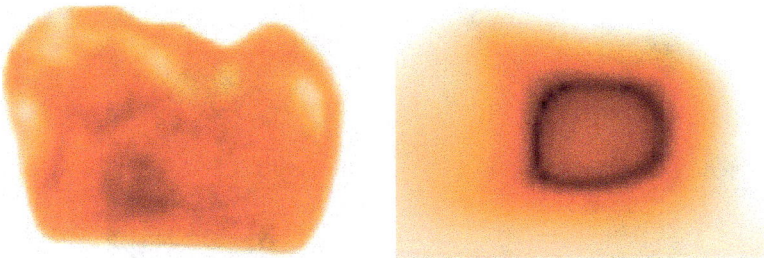

Fig. 8.5.   Confocal (left) and near-field (right) integral THz image of the tooth sample of Fig. 8.4.

(as it can be seen in Fig. 8.5 on the right). An additional advantage of the SNIM is the absence of the ionizing effects of X-rays, especially with repeating tests. There are still challenges for the SNIM applications. In particular, certain biological tissues (*e.g.* enamel) cannot be imaged since the dentin is almost opaque in the near-field infrared. THz pulse imaging in the far-infrared has been more successful for a thin layer of enamel ($100\,\mu$m) (Crowley, 2003) obtaining a higher attenuation of the THz in carious enamel as compared with the healthy enamel. In confocal imaging geometry, the tooth cannot be spatially resolved and the image is strongly blurred due to diffraction. In contrast, the enamel and the dentin regions

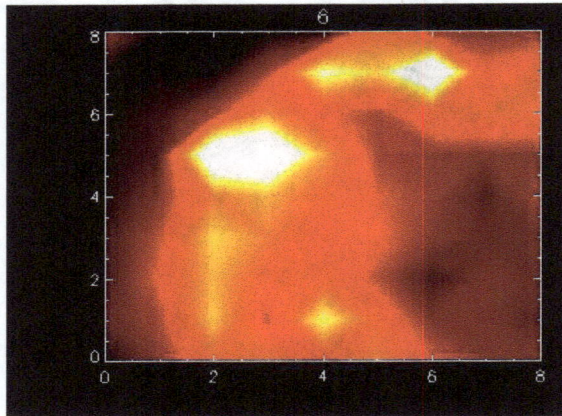

Fig. 8.6.   (a) Spectral near-field image of the lesion area for wavenumber (Lennard-Jones, 1924).

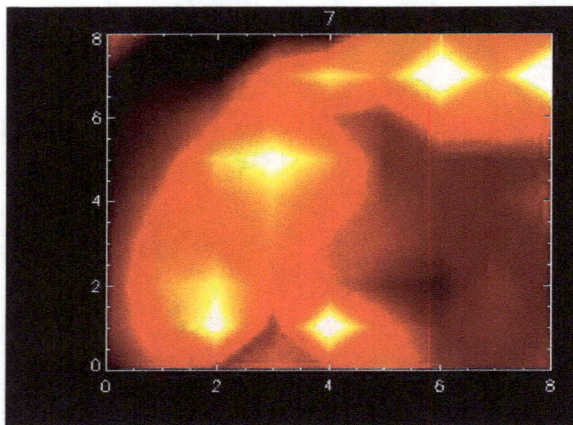

Fig. 8.6.   (b) Spectral near-field image of the lesion area for wavenumber (Tuckerman *et al.*, 1992).

of the tooth sample as well as the buried caries lesions along with the other structural irregularities and defects are spatially resolved in the near-field image. Figures 8.6(a) and 8.6(b) show several images of the area with lesions for different wavenumbers. The lesion areas correspond to the higher attenuation spot in the images.

The capabilities of SNIM presented by the authors were tested on leaves and teeth, however, with the depth of penetration of more than 1 mm and the received resolution, the results of the tests are applicable to other types of biological tissues in the light of defense and security applications. Such applications may include skin ailments (*e.g.* burns) and bone injuries (*e.g.* fractures in bones). At the same time, for example, dentistry applications are also useful for military personnel for field and hospital usage.

8.4. THz characterization of biological and inorganic material (hydrated and anhydrous)

As it was already pointed out, the characterization of different forms of organic and inorganic materials is of great importance to science and technology, including specific aspects of identification of drugs, biological tissues, explosives (that may be in a hydrated or anhydrous form), etc. Water content poses difficulties for successful identification of the material structure by THz radiation. However, as we have seen, biological tissues and hydrated forms of inorganic substances still may be investigated by THz waves.

With many applications of THz identification, water content is one of the crucial issues setting limits or even impeding characterization of materials' properties under investigation. While water itself is an obstacle for THz waves, the presence of water content does not necessarily make measurements of the materials' parameters impossible. Solids that may or may not contain water change their crystal structure while transforming from an anhydrous to a hydrated form with a change of the absorption coefficient. Thus, one of the main goals is to determine, *e.g.* a change of the absorption coefficient for the THz probing beam with the amount of water content and the nature of the crystal or molecular structure. With experimental work revealing the above dependencies, however, the complexity of the dependence calls for mathematical methods to establish the degree of stochasticity involved and calculate the probability of the model that represents the investigated material.

Let us consider the absorption of electromagnetic radiation at different frequencies: *e.g.* in the infrared region, absorption corresponds to motion of nuclei in the molecule. More specifically, near- and mid-infrared relate to stretching or bending motions of individual bonds in the molecule that imply small mass (of individual molecules) and strong bonds. Far-infrared and THz regions ($\sim$0.3–10 THz) involve larger masses and weaker bonds. Different molecules and molecular structures (or different atoms and crystal structures) vibrate differently at different probing frequencies. Thus, a molecular/crystal structure may be characterized by different vibrational modes. The modes may be identified through their frequencies. Further, using spectroscopy, absorption information from spectra may be received by high-resolution measurements. THz radiation interacts with the low-frequency internal vibrations of molecules that involve weak bonds of hydrogen. As a result, the absorption spectra that correspond to the above vibrations may be received. The THz spectroscopy provides unique capabilities to determine the above weak-bond frequencies, which is more difficult or impossible with other methods, *e.g.* by using visible light or IR spectroscopy. The unique information opens new possibilities of THz application of materials containing water. For biological tissue characterization, THz radiation is also safer than X-Ray.

The absorption of materials with water content is different from that of anhydrous materials. Although the absorption coefficient of water in the THz region is significantly smaller than, *e.g.* in the infrared range, its magnitude is still high for many biological samples and in some other cases. Water contains polar molecules that are in dynamic interactions with one another through hydrogen bonds that constitute a whole structure rather than separate entities. Thus, the resulting spectrum is influenced by the interaction among the molecules yielding resonance frequencies. Molecular dynamics (MD) and statistical mechanics are used to determine molecular structure dynamic characteristics on the macro scale. In order to investigate qualities of aqueous bonds on the atomic scale, a number of models have been suggested.

8.4.1. *Modeling of water intermolecular structure*

Intermolecular structure of water constitutes hydrogen bonding. This type of bonding is characterized by the presence of partially-negative oxygen and a partially-positive hydrogen atoms being influenced by neighboring atoms. The water molecules are not stable with constantly breaking bonds. Aqueous structure exhibits a low-potential energy configuration with the hydrogen bond of 0.17 nm, which is longer than a normal and O–H bond of 0.1 nm that is weaker than a normal bond. The lifetime of such a configuration is about several picoseconds. Following the breaks of bonds and disintegration, the molecules again form tetrahedrons, thus enabling water to remain structurally stable. The structural stability also specifies absorption peaks significant for water content influence on THz penetration of hydrated materials. The most popular models of water molecular interaction present three interaction points centered around the atomic nuclei. The model assumes a rigid structure (monomer). The distinctions arising from several modifications of the model are in the length of O–H bond and the HOH angle (Gordon and Johnson, 2006). Lennard-Jones's expression (Lennard-Jones, 1924) gives the potential of intermolecular interactions:

$$V_{inter} = \sum \{4\varepsilon((\sigma/R_0)^{12} - (\sigma/R_0)^6)\} + \sum q_i q_j e^2/R_{ij}, \qquad (8.3)$$

where R_0 is a distance between oxygen atoms; ε and σ are the Lennard-Jones parameters for oxygen atom pairs; R_{ij} is the distance between atoms i and j; e is the charge of an electron and q_i is a fractional charge force due to atom i, σ is the finite distance at which the inter-particle potential is zero. The term under $()^{12}$ represents repulsive forces (Pauli repulsion at short ranges due to overlapping electron orbitals). The term under $()^6$ represents the attracting term (due to attraction at long ranges which is van der Waals force or dispersion force). The Lennard-Jones expression is an approximation. The lowest energy of an infinite number of atoms described by the Lennard-Jones' approximation, corresponds to a hexagonal close-packing. As the temperature rises, the lowest energy will correspond first to cubic close-packing, and then to the liquid form. The Lennard-Jones' parameters for interactions between oxygen atoms based on the SPCE model (Starr *et al.*, 2000) are: $\varepsilon \approx 0.1554\,\mathrm{kcal/mol}$ and $\sigma \approx 0.317\,\mathrm{nm}$. The fraction charge for oxygen is assumed to be -0.848 and for hydrogen 0.424.

A Monte Carlo simulation may be applied to a water special fragment containing as many as several hundred molecules. The density of the fragment should approximately correspond to the density of the water under normal conditions. The pressure and temperature should be at normal laboratory conditions, which are the pressure of $\sim 1\,\mathrm{atm}$ and the temperature of 298 K. The weak coupling algorithm commonly known as Berendsen thermostat is necessary to achieve normal kinetic energies of the water molecules corresponding to the above-indicated temperature and pressure. For the special water fragment, the Cartesian component of the dipole moment,

$$M_\gamma = \sum e_i \gamma_i, \qquad (8.4)$$

where e_i is the electric charge of the atom i, $\sum e_i \gamma_i$ sums up all atoms with the Cartisian coordinate γ. The total charge is zero (for a neutral molecule).

Further, the dipole absorption spectrum is determined by comparing the spectrum at equilibrium to that when an external electromagnetic field is applied. The linear response for the absorption cross-section is:

$$\alpha = (2\pi\omega^2/3nc) \int_{-\infty}^{+\infty} dt e^{-i\omega t} \langle \mathbf{M}(0) \cdot \mathbf{M}(t) \rangle, \qquad (8.5)$$

where n is the refraction index of the medium, $c =$ speed of light, $\omega = 2\pi\vartheta$ (circular frequency), $M(t)$ is the dipole moment, and $\langle \mathbf{M}(0) \cdot \mathbf{M}(t) \rangle$ is the time correlation average for the dipole moment. The average of the dipole moment temporary correlation function:

$$\langle \mathbf{M}(0) \cdot \mathbf{M}(t) \rangle = \lim_{\tau \to \infty} \sum \gamma = x, y, z \left| \int dt e^{-i\omega t} M_\gamma(t) \right|^2 . \qquad (8.6)$$

τ and t come from the cumulative distribution function $\{X_t\}$ at times $t_1 + \tau \ldots t_k + \tau$. The linear response α, however is a stationary process. The application of SPCE model yields absorption spectrum as shown in Fig. 8.7.

The spectra in Fig. 8.7 contain noise. The noise reduction is performed under assumption that the spectrum has a Gaussian shape or is a sum of Gaussian shapes. In this case, the Gaussian distribution is used as a model

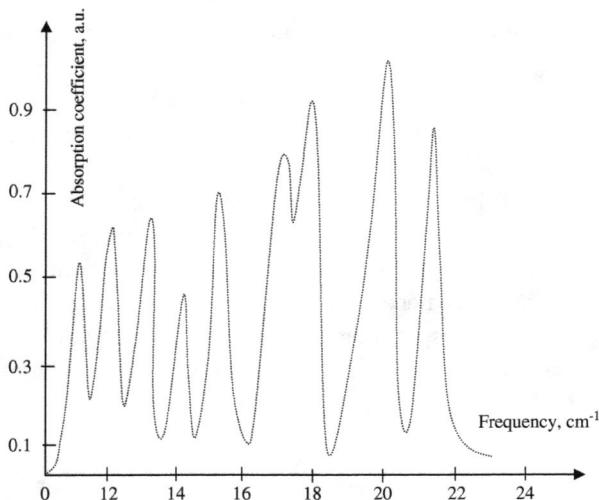

Fig. 8.7. Water absorption spectrum. Molecular Dynamical (MD) (Tuckerman *et al.*, 1992) simulation with SPCE model.

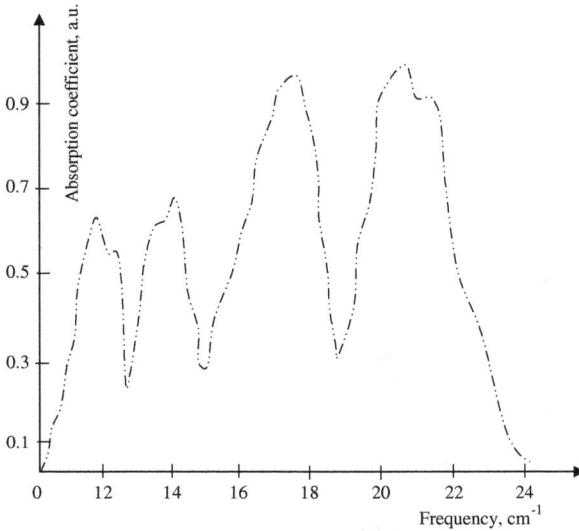

Fig. 8.8. Normalized absorption spectra of water in THz range.

for a complex phenomenon. For the Gaussian distribution:

$$f(x) = ae^{-\frac{(x-\mu)^2}{2\sigma^2}}, \tag{8.7}$$

where a is a constant, σ^2 is used as a window parameter (gives a narrow band of frequency to screen unwanted parts of the spectra). Taking the special frequency $\vartheta = \frac{1}{\lambda}$ of $0.3\,\mathrm{cm}^{-1}$, the spectra becomes the spectra shown in Fig. 8.8.

The spectra in Fig. 8.8 contains noise that may be eliminated by comparison of the received spectra with the theoretical model (SPCE) and the prediction time series model (Sokolnikov, 2006) (Fig. 8.9).

8.4.1.1. *THz spectroscopy of hydrated and anhydrous substances*

Measurements of the absorption coefficient of dry (anhydrous) explosive RDX were performed in the range of 0 to 3 THz (see Fig. 8.10, the lower curve). Scattering increases with the increase of the frequency of the THz probing beam. Only noise and random disturbances cause fluctuations of the curve that otherwise is close to a straight line. If the RDX contains water molecules, the absorption and scattering pattern change as we see in the curve above the anhydrous form line in Fig. 8.10. There are

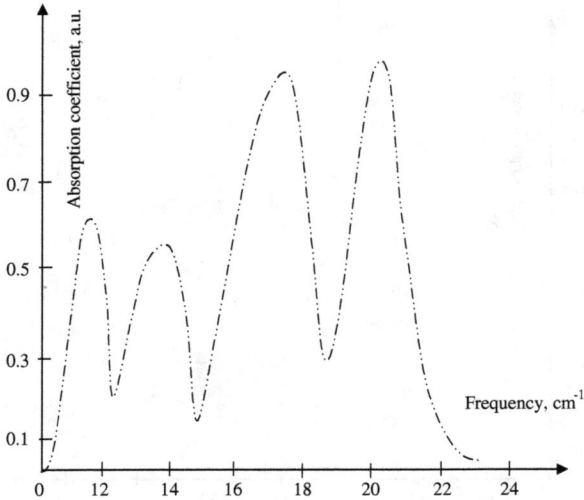

Fig. 8.9.   Processed by time series prediction absorption spectra (without noise).

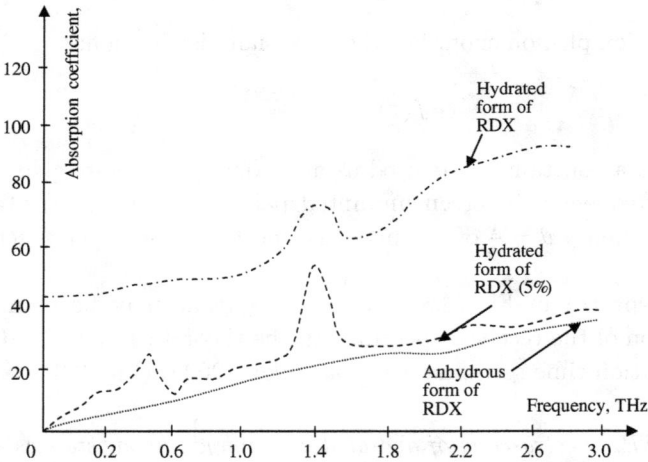

Fig. 8.10.   Frequency dependence of the absorption coefficient of RDX explosive for anhydrous and hydrated forms.

several resonance peaks that have frequencies close to 1.4–1.6 THz. Similar resonance peaks were found in a number of other hydrated forms, *e.g.* theophylline and D-gluocose (Liu *et al.*, 2007). From the previous section, we know that the absorption pattern is determined by the dipole moment

M_γ (3). However, in the case of a hydrated form, the resonance peaks shift as a result of the complex combination of water intermolecular structure and the one of the solid (as in the case of RDX). The increase in the water content decreases the sharpness of the resonance peaks.

THz-TDS is generally easy to use for identifying substances in sealed containers transparent to THz radiation. The deleted signal is absorbed by packaging through the mechanism that discriminates the different materials of which the enclosed substances and container are made. For example, in the case of a crystalline material inside the container, the specific decay signal is emitted coherently by the inside sample due to excitation of collective vibrational modes of the crystal structure by the ultrashort (pico seconds or fractures of) broadband THz pulses. The difference between the absorption signal from the packaging, anhydrous and hydrated pharmaceutical can be seen in Fig. 8.11 which depicts the spectra of a common pharmaceutical, aspirin in two forms: the usual, anhydrous and with some water content (the hydrated form). Below, in Fig. 8.11, there is another curve that presents the packaging influence on the drug identification. It does not screen off the responses from the hydrated and anhydrous aspirin, although the thickness of the polythene

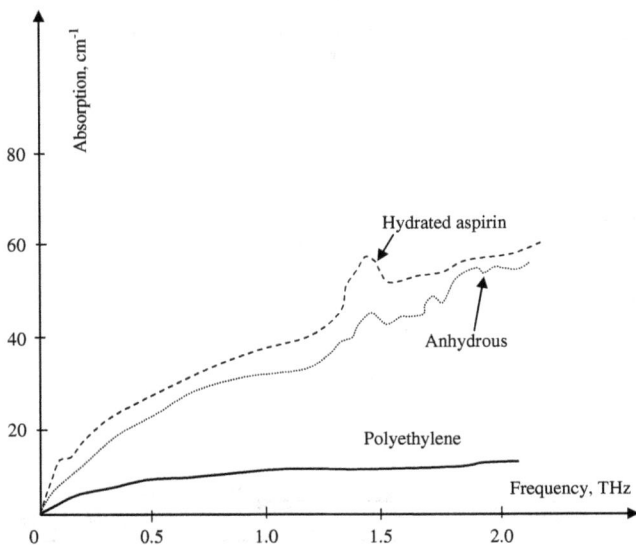

Fig. 8.11.   Absorption spectra of hydrated and anhydrous forms of aspirin and the packaging.

envelope has substantial significance. In the described case, the polyethylene 1 mm-wrapping was used. With the increase of thickness and usage of other materials (*e.g.* paper) the absorption may increase dramatically, which should be taken into account for practical designs of THz identification set-ups. The characteristic resonance peak at $f = 1.6$ THz of the hydrated form almost coincides with the anhydrous crystal's response in form but differs in intensity. Also, the characteristic increase of absorption with frequency is present. It is caused by scattering and is stronger for crystal structures. The scattering, however, does not change the identification signatures — the form of the curves with resonant peaks. The next graph (Fig. 8.12) shows a distinctly different pattern of absorption for lactose (which is similar to that of cocaine and morphine) with several resonance peaks and the packaging response below.

The resonance peaks constitute the recognition (or signature) pattern of identification. The water molecules superimpose their absorption spectra on those of anhydrous lactose; nevertheless, it is still possible to identify the crystalline pharmaceutical. In the THz range, transitions between vibrational states of the identified crystal come for the most part from the characteristic crystal lattice absorption that becomes more pronounced with the quality of the crystalline pharmaceutical.

Another example of changes introduced by water content is given in Fig. 8.13 for succinic acid, which is a solid that forms colorless, odorless

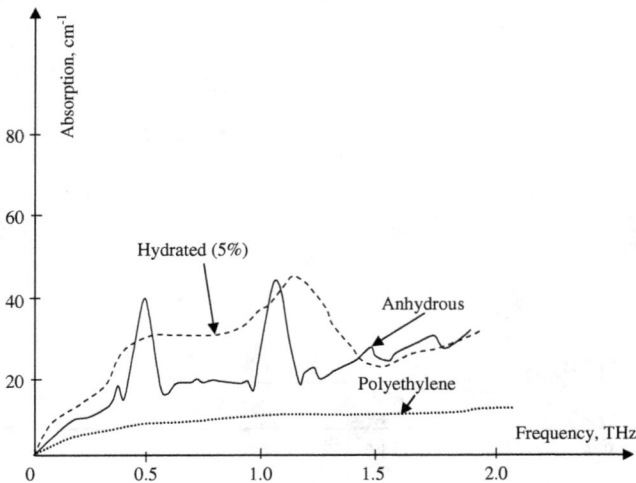

Fig. 8.12. Comparison spectra of the hydrated and anhydrous forms of lactose.

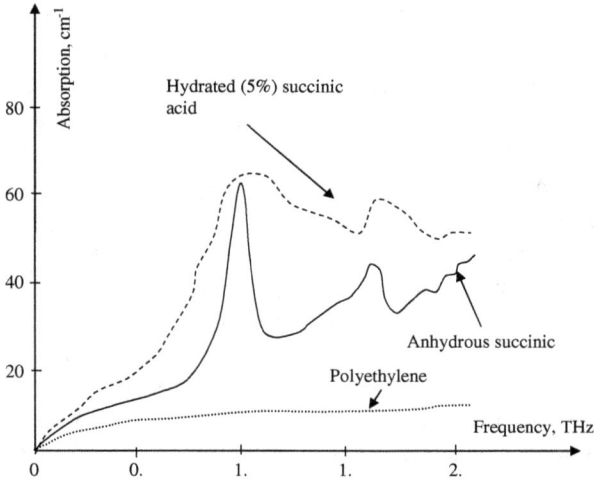

Fig. 8.13. Spectra of the hydrated and anhydrous forms of succinic acid.

crystals at room temperature. The acid is combustible and corrosive, capable of causing burns. There are two distinct resonance peaks that are characteristic of many drugs (pharmaceuticals) and explosives and are signatures for their identification. Both peaks happen within the range of 0–2 THz accessible for most THz-TDS set-ups. Again, the thin layer of packaging does not obstruct the THz wave, although it reduces sensitivity.

8.4.1.2. *Conclusions for the examples and theory*

The measurements of the particular substances provided above confirm the main points of the water intermolecular structure modeling described in "Modeling of water intermolecular structure". We have seen that many substances of interest, especially of crystalline nature exhibit intrinsic resonance peaks due to molecular vibration, stretching and bending of the crystal or molecular structures, allowing THz sensing to have a potential for noninvasive and label-free detection of materials. Pharmaceuticals and explosives, in particular, are likely to be investigated by THz waves with water content not being a fundamental obstacle for the identification. In particular, in biological tissues (where water is almost invariably present), THz radiation interacts with molecules by exciting low-frequency molecular vibrations that involve groups connected by weak hydrogen bonds (see "Modeling of water intermolecular structure") and van der

Waals hydrophobic interactions. These weak bonds reveal signature data related to pharmaceutical substances and biological tissues thus enabling to receive images of living (and not necessarily dead) biological structures. In examples provided above, calculations of refraction and absorption indices were performed (the latter were given in graphs) that were not substantially changed by water content addition. Studies also demonstrated that a low-percentage of water content in intrinsically anhydrous materials does not change the resonance frequencies and the resonance peaks only shift in amplitude with some flattening of the curves (Figs. 8.10–8.13). It is also important to mention that water content is an extricable part of a biological tissue that yields THz images containing information about DNA structural elements of substance. The lower absorption at the lower frequencies of the THz range (*i.e.* lower than 1.0 THz) compared to the IR range allowed characterization of biological materials by THz Fourier transform spectroscopy (Globus *et al.*, 2006).

References

Crowley, DA, C Longbottom, BE Cole, CM Ciesla, D Arnone, VP Wallace and M Pepper (2003). Terahertz pulse imaging: A pilot study of potential applications in dentistry. *Caries Research*, 37, 352–359.

Globus, T, T Khromova, D Woolard and A Samuels (2006). THz resonance spectra of *Bacillus Subtilis* cells and spores in PE pellets and dilute water solutions. Terahertz for military and security applications IV. *Proc. SPIE*, 6212, 1–12.

Gordon, TH and ME Johnson (2006). Tetrahedral structure of chains for liquid water. *Proc. Acad. Sci. USA*, 103, 7073–7977.

Lennard-Jones, JE (1924). On the determination of molecular fields. *P. Roy. Soc. Lond. Series A, Mathematical and Physical Sciences*, 106(738), 463–477.

Liu, HB, Y Chen and X-C Zhang (2007). Characterization of anhydrous and hydrated pharmaceutical materials with THz time-domain spectroscopy. *J. Pharm. Sci.*, 96(4).

Schade, U, K Holldack, MC Martin and D Fried (2005). THz near-field imaging of biological tissues employing synchrotron radiation. *Proc. SPIE*, 5725, 46.

Sokolnikov, A (2005). Resonance amplification of the probing signals in optical coherence tomography (OCT). *Proc. SPIE*, 5881.

Sokolnikov, A (2006). Adaptive non-intrusive terahertz identification. *Proc. SPIE*, 6212.

Starr, FW, JK Nielsen and HE Stanley (2000). Hydrogen-bond dynamics for the extended simple point-charge model of water. *Phys. Rev.*, 62(1).

Tuckerman, ME, BJ Berne and GJ Martyna (1992). Reversible multiple time scale molecular dynamics. *J. Chem. Phys.*, 97(3).

Index

217

www.ingramcontent.com/pod-product-compliance
Lightning Source LLC
Chambersburg PA
CBHW050559190326
41458CB00007B/2100